Praise for

Mistakes Were Made (but Not by Me)

AN AMAZON BEST BUSINESS BOOK OF 2008

"Anecdote-rich ... A ramble through the evasive tactics we employ when we've done something wrong and don't want to face up to it ... By turns entertaining, illuminating, and — when you recognize your-self in the stories it tells — mortifying."
  — *Wall Street Journal*

"A revelatory study of how lovers, lawyers, doctors, politicians — and all of us — pull the wool over our own eyes ... Thanks, in part, to the scientific evidence it provides and the charm of its down-to-earth, commonsensical tone, *Mistakes Were Made* is convincing. Reading it, we recognize the behavior of our leaders, our loved ones, and — if we're honest — ourselves, and some of the more perplexing myster-ies of human nature begin to seem a little clearer. By the book's end, we're far more attuned to the ways in which we avoid admitting our missteps, and intensely aware of how much our own (and everyone's) lives would improve if we — and those who govern and lead us — understood the power and value of simply saying, 'I made a mistake. I'm sorry.'"
  — Francine Prose, *O, the Oprah Magazine*

"This book is charming and delightful. But mainly, it's just damn smart . . . Every page sparkles with sharp insight and keen observation. Mistakes were made — but not in this book!"
— Daniel Gilbert, author of *Stumbling on Happiness*

"A dream team of two of psychology's greatest communicators . . . A fascinating exploration of our astonishing powers of self-justification."
— David G. Myers, author of *Intuition: Its Powers and Perils*

"This eye-opener of a book is essential reading, not because we've all made mistakes — certainly not! — but because we've all been victims of mistakes made by others. Why do these people behave so badly? Tavris and Aronson's explanation is illuminating, entertaining, based on solid science, and highly relevant to our public and private lives."
— Judith Rich Harris, author of *The Nurture Assumption* and *No Two Alike*

"It is hard to think of a better — or more readable — guide to the mind's most devilish tricks."
— David Callahan, author of *The Cheating Culture*

"One of the most needed and important books for our time."
— Warren Bennis, author of *On Becoming a Leader*

"If this book doesn't change the way we think about our mistakes, then we're all doomed."
— Michael Shermer, editor of *Skeptic* magazine and author of *Why People Believe Weird Things*

"Tavris and Aronson have combined their formidable skills to produce a gleaming model of social insight and scientific engagement. Make no mistake, you need to read this book."

— Robert B. Cialdini, author of *Influence: Science and Practice*

"Combining far-ranging scholarship with lucid, witty prose, Tavris and Aronson illuminate many of the mysteries of human behavior — why hypocrites never see their own hypocrisy, why couples so often misremember their shared history, why many people persist in courses of action that lead straight into quicksand. A delight to read, with surprising revelations in every chapter."

— Elizabeth Loftus, author of *Eyewitness Testimony* and
distinguished professor at the University of California, Irvine

"A pathbreaking book that could change forever how leaders think about the decisions they make. Crackles with new insights and understanding. A must-read!"

— Burt Nanus, coauthor of *Leaders: Strategies for Taking Charge*

MISTAKES WERE MADE
(BUT NOT BY *ME*)

Why We Justify Foolish Beliefs,
Bad Decisions, and Hurtful Acts

Carol Tavris and
Elliot Aronson

MARINER BOOKS
Houghton Mifflin Harcourt
Boston   New York

*For Leon Festinger, creator of the theory of cognitive
dissonance, whose ingenuity inspired this book*

For information about permission to reproduce selections from
this book, write to trade.permissions@hmhco.com or to Permissions,
Houghton Mifflin Harcourt Publishing Company, 3 Park Avenue,
19th Floor, New York, New York 10016.

hmhbooks.com

Library of Congress Cataloging-in-Publication Data is available.

ISBN 978-0-358-32961-9

Printed in the United States of America

DOC 10 9 8 7 6 5 4 3 2 1

"Frank and Debra" extract from Andrew Christensen and Neil S. Jacobson's
Reconcilable Differences is © 2000 Guilford Press and is reprinted with
permission of Guilford Press.

We are all capable of believing things which we know to be untrue, and then, when we are finally proved wrong, impudently twisting the facts so as to show that we were right. Intellectually, it is possible to carry on this process for an indefinite time: the only check on it is that sooner or later a false belief bumps up against solid reality, usually on a battlefield.

— *George Orwell, 1946*

I see no reason why I should be consciously wrong today because I was unconsciously wrong yesterday.

— *Supreme Court Justice Robert H. Jackson, 1948*

CONTENTS

PREFACE TO
THE REVISED EDITIONS

When the first edition of this book was published, in 2007, the country had already become polarized by the war in Iraq. Although Democrats and Republicans were initially equally likely to support George W. Bush's decision to invade, believing that Saddam Hussein was developing weapons of mass destruction, it soon became clear that he wasn't, and none were ever found. WMDs had vanished, but not political polarization, which we saw for ourselves in the reviewers of our book on Amazon.

Many conservatives were (and some still are) deeply annoyed by their perception that we were bashing Bush unfairly. One, who titled his review "Almost Great" and gave *Mistakes Were Made* three stars, said the book would have been truly great if we hadn't spent so much damned time trying to impose our political views on the reader and ignoring the mistakes and bad decisions that Democrats made. Any future edition, he advised, should delete all the "Bush lied" examples so it didn't seem like there was one on every fourth page.

Then we found a rebuttal review headed "Truly Great!" and

giving the book five stars. This isn't a book about politics alone, this reviewer said, but about all aspects of human behavior. She found it extremely balanced, noting it discussed the mistakes, self-justifications, and delusions of members of both parties — for example, Lyndon Johnson's inability to get out of Vietnam was compared to Bush's determination to "stay the course" in Iraq.

For reasons that will be clear as you read this book, we enjoyed the second of these two Amazon reviews much more than the first. What a brilliant, astute reader, we thought, obviously so well informed! Whereas the first reviewer was completely muddled. Biased? *Us?* Don't be absurd! Why, we bent over backward to be fair! A Bush-lied example on every fourth page and we didn't have a bad word for Democrats? Didn't this reader see our criticism of LBJ, whom we called a "master of self-justification"? How did he miss the Republicans we praised? And how did he misunderstand our main point, that George Bush was *not* intentionally lying to the American public about Saddam Hussein's alleged weapons of mass destruction but doing something all leaders and the rest of us do: lying to himself to justify a decision he had already made? And besides, we said, warming to our own defense, Bush was president when we began writing this book, and the costly war was dividing the nation. Its consequences are with us today, in the continuing warfare and chaos in the Middle East. What other example could have been as powerful or important an opening story?

Then, after reveling in our spasm of self-justification in response to the first reviewer, we had to face the dreaded question: "Wait a minute — are we right, or are we merely justifying ourselves? What if — horrors! — he has a point?" As human beings, the two of us are not immune to the pitfalls of thinking that we describe in our own book. No human being can live without biases, and we have ours. But we wrote this book with the goal of understanding them and shining a light on their operation in all corners of people's lives, including our own.

In the years since this book first appeared, readers, reviewers, neighbors, and friends have sent us comments, studies, and personal stories. Professionals in fields as different as dentistry, engineering, education, and nutrition urged us to add chapters on *their* experiences with recalcitrant colleagues who refused to pay attention to the data. Friends in England and Australia formed the Mistakes Were Made Irregulars to let us know who was using this iconic phrase in their countries.

We realized that a revision could easily be twice as long as the original without being twice as informative. For the second edition (2015), we updated the research and offered examples of attempts by organizations to correct mistakes and end harmful practices (for instance, in criminal prosecutions, methods of interrogation, hospital policies, and conflicts of interest in science). Tragically, but not surprisingly for anyone who reads this book, there have not been nearly enough of those systematic corrections, and in some areas, deeply felt but incorrect beliefs, such as those held by people who oppose vaccinating their children, have become even more entrenched. We made a major change in chapter 8 by addressing an issue we had intentionally avoided the first time around: the problems that arise for people who cannot justify their mistakes, harmful actions, or bad decisions and who, as a result, suffer PTSD, guilt, remorse, and sleepless nights for far too long. There we offered research and insights that might help people find a path between mindless self-justification and merciless self-flagellation, a path worth struggling to discover.

And then, not long after the second edition appeared, Donald Trump was elected president of the United States, immediately exacerbating the political, ethnic, racial, and demographic tensions that had been growing for decades. Of course, political polarization between left and right, progressive and traditional, urban and rural, has existed throughout history and is still found all over the planet, with each side seeing the world through its preferred lens. But the Trump phenomenon is unique in American history, because Trump intentionally violated the rules, norms, protocols, and procedures of government — actions that his supporters applauded, his adversaries condemned, and many of his former opponents came to endorse. Whether or not Trump is in office as you read this, Americans will long be facing the moral, emotional, and political residue of his presidency.

It seems like eons since Republican nominee Bob Dole described Bill Clinton as "my opponent, not my enemy," but in fact he made that civilized remark in 1996. How quaint it now seems in contrast to Donald Trump, who regards his opponents (or people who simply disagree with him) as treasonous, disloyal rats and foes. In our new concluding chapter, therefore, we closely examine the process by which Trump, his administration, and his supporters fostered that view, with devastating consequences for our democracy. We wrote this chapter in the hope that once we understand the slow but pernicious shift in thinking from opponent to enemy, we can begin to find our way back.

— *Carol Tavris and Elliot Aronson, 2020*

INTRODUCTION

Knaves, Fools, Villains, and Hypocrites:
How Do They Live with Themselves?

Mistakes were quite possibly made by the administrations in which I served.

— *Henry Kissinger, responding to charges that he committed war crimes in his role in the United States' actions in Vietnam, Cambodia, and South America in the 1970s*

If, in hindsight, we also discover that mistakes may have been made . . . I am deeply sorry.

— *Cardinal Edward Egan of New York (referring to the bishops who failed to deal with child molesters among the Catholic clergy)*

We know mistakes were made.

— *Jamie Dimon, CEO of JPMorgan Chase (referring to enormous bonuses paid to the company's executives after the government bailout had kept them from bankruptcy)*

> Mistakes were made in communicating to the public and customers about the ingredients in our French fries and hash browns.
>
> — *McDonald's (apologizing to vegetarians for failing to inform them that the "natural flavoring" in its potatoes contained beef byproducts)*

As fallible human beings, all of us share the impulse to justify ourselves and avoid taking responsibility for actions that turn out to be harmful, immoral, or stupid. Most of us will never be in a position to make decisions affecting the lives and deaths of millions of people, but whether the consequences of our mistakes are trivial or tragic, on a small scale or a national canvas, most of us find it difficult if not impossible to say "I was wrong; I made a terrible mistake." The higher the stakes — emotional, financial, moral — the greater the difficulty.

It goes further than that. Most people, when directly confronted by evidence that they are wrong, do not change their point of view or plan of action but justify it even more tenaciously. Politicians, of course, offer the most visible and, often, most tragic examples of this practice. We began writing the first edition of this book during the presidency of George W. Bush, a man whose mental armor of self-justification could not be pierced by even the most irrefutable evidence. Bush was wrong in his claim that Saddam Hussein had weapons of mass destruction; he was wrong in stating that Saddam was linked with al-Qaeda; he was wrong in his prediction that Iraqis would be dancing joyfully in the streets at the arrival of American soldiers; he was wrong in his assurance that the conflict would be over quickly; he was wrong in his gross underestimate of the human and financial costs of the war; and he was most famously wrong in his speech six weeks after the invasion began when he announced (under a banner reading MISSION ACCOMPLISHED) that "major combat operations in Iraq have ended."

Commentators from the right and left began calling on Bush to admit he had been mistaken, but Bush merely found new justifications for the war: he was getting rid of a "very bad guy," fighting terrorists, promoting peace in the Middle East, bringing democracy to Iraq, increasing American security, and finishing "the task [our troops] gave their lives for." In the midterm elections of 2006, which most political observers regarded as a referendum on the war, the Republican Party lost both houses of Congress; a report issued shortly thereafter by sixteen American intelligence agencies announced that the occupation of Iraq had actually *increased* Islamic radicalism and the risk of terrorism. Yet Bush said to a delegation of conservative columnists, "I've never been more convinced that the decisions I made are the right decisions."[1]

George Bush was not the first nor will he be the last politician to justify decisions that were based on incorrect premises or that had disastrous consequences. Lyndon Johnson would not heed the advisers who repeatedly told him the war in Vietnam was unwinnable, and he sacrificed his presidency because of his self-justifying certainty that all of Asia would "go Communist" if America withdrew. When politicians' backs are against the wall, they may reluctantly acknowledge *error* but not their responsibility for it. The phrase "Mistakes were made" is such a glaring effort to absolve oneself of culpability that it has become a national joke — what the political journalist Bill Schneider called the "past exonerative" tense. "Oh, all right, mistakes were made, but not by me, by someone else, someone who shall remain nameless."[2] When Henry Kissinger said that the administration in which he'd served may have made mistakes, he was sidestepping the fact that as national security adviser and secretary of state (simultaneously), he essentially *was* the administration. This self-justification allowed him to accept the Nobel Peace Prize with a straight face and a clear conscience.

We look at the behavior of politicians with amusement or alarm or horror, but what they do is no different in kind, though certainly in consequence, from what most of us have done at one time or another in our private lives. We stay in an unhappy relationship or one that is merely going nowhere because, after all, we invested so much time in making it work. We stay in a deadening job way too long because we look for all the reasons to justify staying and are unable to clearly assess the benefits of leaving. We buy a lemon of a car because it looks gorgeous, spend thousands of dollars to keep the damn thing running, and then spend even more to justify that investment. We self-righteously create a rift with a friend or relative over some real or imagined slight yet see ourselves as the pursuers of peace — if only the other side would apologize and make amends.

Self-justification is not the same thing as lying or making excuses. Obviously, people will lie or invent fanciful stories to duck the fury of a lover, parent, or employer; to keep from being sued or sent to prison; to avoid losing face; to avoid losing a job; to stay in power. But there is a big difference between a guilty man telling the public something he knows is untrue ("I did not have sex with that woman"; "I am not a crook") and that man persuading himself that he did a good thing. In the former situation, he is lying and knows he is lying to save his own skin. In the latter, he is lying to himself. That is why self-justification is more powerful and more dangerous than the explicit lie. It allows people to convince themselves that what they did was the best thing they could have done. In fact, come to think of it, it was the right thing. "There was nothing else I could have done." "Actually, it was a brilliant solution to the problem." "I was doing the best for the nation." "Those bastards deserved what they got." "I'm entitled."

Self-justification minimizes our mistakes and bad decisions; it also explains why everyone can recognize a hypocrite in action except the hypocrite. It allows us to create a distinction between our moral lapses and someone else's and blur the discrepancy between our actions and our moral convictions. As a character in Aldous Huxley's novel *Point Counter Point* says, "I don't believe there's such a thing as a conscious hypocrite." It seems unlikely that former Speaker of the House and Republican strategist Newt Gingrich said to himself, "My, what a hypocrite I am. There I was, all riled up about Bill Clinton's sexual affair, while I was having an extramarital affair of my own right here in town." Similarly, the prominent evangelist Ted Haggard seemed oblivious to the hypocrisy of publicly fulminating against homosexuality while enjoying his own sexual relationship with a male prostitute.

In the same way, we each draw our own moral lines and justify them. For example, have you ever done a little finessing of expenses on income taxes? That probably compensates for the legitimate expenses you forgot about, and besides, you'd be a fool not to, considering that everybody else does it. Did you fail to report some extra cash income? You're entitled, given all the money that the government wastes on pork-barrel projects and programs you detest. Have you been texting, writing personal e-mails, and shopping online at your office when you should have been tending to business? Those are perks of the job, and besides, it's your own form of protest against those stupid company rules, plus your boss doesn't appreciate all the extra work you do.

Gordon Marino, a professor of philosophy and ethics, was staying in a hotel when his pen slipped out of his jacket and left an ink spot on the silk bedspread. He decided he would tell the manager, but he was tired and did not want to pay for the damage. That evening he went out with some friends and asked their advice. "One of them told me to stop with the moral fanaticism," Marino said. "He argued, 'The management expects such accidents and builds their cost into the price of the rooms.' It did not take long to persuade me that there was no need to trouble the manager. I reasoned that if I had spilled this ink in a family-owned bed-and-breakfast, then I would have immediately reported the accident, but that this was a chain hotel, and yadda yadda yadda went the hoodwinking process. I did leave a note at the front desk about the spot when I checked out."[3]

But, you say, all those justifications are true! Hotel-room charges do include the costs of repairs caused by clumsy guests! The government does waste money! My company probably wouldn't mind if I spend a little time texting and I do get my work done (eventually)! Whether those claims are true or false is irrelevant. When we cross these lines, we are justifying behavior that we know is wrong precisely so that we can continue to see ourselves as honest people and not criminals or thieves. Whether the behavior in question is a small thing like spilling ink on a hotel bedspread or a big thing like embezzlement, the mechanism of self-justification is the same.

Now, between the conscious lie to fool others and unconscious self-justification to fool ourselves, there's a fascinating gray area patrolled by an unreliable, self-serving historian — memory. Memories are often pruned and shaped with an ego-enhancing bias that blurs the edges of past events, softens culpability, and distorts what really happened. When researchers ask wives what percentage of the housework they do, they say, "Are you kidding? I do almost everything, at least 90 percent." And when they ask husbands the same question, the men say, "I do a lot, actually, about 40 percent." Although the specific numbers differ from couple to couple, the total always exceeds 100 percent by a large margin.[4] It's tempting to conclude that one spouse is lying, but it is more likely that each is remembering in a way that enhances his or her contribution.

Over time, as the self-serving distortions of memory kick in and we forget or misremember past events, we may come to believe our own lies, little by little. We know we did something wrong, but gradually we begin to think it wasn't all our fault, and after all, the situation was complex. We start underestimating our own responsibility, whittling away at it until it is a mere shadow of its former hulking self. Before long, we have persuaded ourselves to believe privately what we said publicly. John Dean, Richard Nixon's White House counsel, the man who blew the whistle on the conspiracy to cover up the illegal activities of the Watergate scandal, explained how this process works:

INTERVIEWER: You mean those who made up the stories were believing their own lies?

DEAN: That's right. If you said it often enough, it would become true. When the press learned of the wire taps on newsmen and White House staffers, for example, and flat denials failed, it was claimed that this was a national-security matter. I'm sure many people believed that the taps *were* for national security; they weren't. That was concocted as a justification after the fact. But when they said it, you understand, they really *believed* it.[5]

Like Nixon, Lyndon Johnson was a master of self-justification. According to his biographer Robert Caro, when Johnson came to believe in something, he would believe in it "totally, with absolute conviction, regardless of previous beliefs, or of the facts in the matter." George Reedy, one of Johnson's aides, said that LBJ "had a remarkable capacity to convince himself that he held the principles he should hold at any given time, and there was something charming about the air of injured innocence with which he would treat anyone who brought forth evidence that he had held other views in the past. It was not an act . . . He had a fantastic capacity to persuade himself that the 'truth' which was convenient for the present was *the truth* and anything that conflicted with it was the prevarication of enemies. He literally willed what was in his mind to become reality."[6] Although Johnson's supporters found this to be a rather charming aspect of the man's character, it might well have been one of the major reasons that Johnson could not extricate the country from the quagmire of Vietnam. A president who justifies his actions to the public might be induced to change them. A president who justifies his actions to himself, believing that he has *the truth,* is impervious to self-correction.

• • •

The Dinka and Nuer tribes of the Sudan have a curious tradition. They extract the permanent front teeth of their children — as many as six bottom teeth and two top teeth — which produces a sunken chin, a collapsed lower lip, and speech impediments. This practice apparently began during a period when tetanus (lockjaw, which causes the jaws to clench together) was widespread. Villagers began pulling out their front teeth and those of their children to make it possible to drink liquids through the gap. The lockjaw epidemic is long past, yet the Dinka and Nuer are still pulling out their children's front teeth.[7] How come?

In the 1840s, a hospital in Vienna was facing a mysterious, terrifying problem: an epidemic of childbed fever was causing the deaths of about 15 percent of the women who delivered babies in one of the hospital's two maternity wards. At the epidemic's peak month, *one-third* of the women who delivered there died, three times the mortality rate of the other maternity ward, which was attended by midwives. Then a Hungarian physician named Ignaz Semmelweis came up with a hypothesis to explain why so many women in his hospital were dying of childbed fever in that one ward: The doctors and medical students who delivered the babies there were going straight from the autopsy rooms to the delivery rooms, and even though no one at the time knew about germs, Semmelweis thought they might be carrying a "morbid poison" on their hands. He instructed his medical students to wash their hands in a chlorine antiseptic solution before going to the maternity ward—and the women stopped dying. These were astonishing, lifesaving results, and yet his colleagues refused to accept the evidence: the lower death rate among Semmelweis's patients.[8] Why didn't they embrace Semmelweis's discovery immediately and thank him effusively for finding the reason for so many unnecessary deaths?

After World War II, Ferdinand Lundberg and Marynia Farnham published the bestseller *Modern Woman: The Lost Sex,* in which they claimed that a woman who achieved in "male spheres of action" might seem to be successful in the "big league," but she paid a big price: "Sacrifice of her most fundamental instinctual strivings. She is not, in sober reality, temperamentally suited to this sort of rough and tumble competition, and it damages her, particularly in her own feelings." And it even makes her frigid: "Challenging men on every hand, refusing any longer to play even a relatively submissive role, multitudes of women found their capacity for sexual gratification dwindling."[9] In the ensuing decade, Dr. Farnham, who earned her MD from the University of Minnesota and did postgraduate work at Harvard Medical School, made a career out of telling women not to have careers. Wasn't she worried about becoming frigid and damaging her own fundamental instinctual strivings?

The sheriff's department in Kern County, California, arrested a retired high-school principal, Patrick Dunn, on suspicion of murdering his wife. The officers had interviewed two people who gave conflicting information. One was a woman who had no criminal record and no personal incentive to lie about the suspect and who had calendars and her boss to back up her account of events; her story supported Dunn's innocence. The other was a career criminal facing six years in prison who had agreed to testify against Dunn as part of a deal with prosecutors and who offered nothing beyond his own word to support his statement; his story suggested Dunn's guilt. The detectives had a choice: believe in the woman (and therefore Dunn's innocence) or the criminal (and therefore Dunn's guilt). They chose the criminal.[10] Why?

By understanding the inner workings of self-justification, we can answer these questions and make sense of dozens of other things people do that otherwise seem unfathomable or crazy. We can answer the question so many people ask when they look at ruthless dictators, greedy corporate CEOs, religious zealots who murder in the name of God, priests who molest children, or family members who cheat their relatives out of inheritances: How in the world can they *live* with themselves? The answer is: exactly the way the rest of us do.

Self-justification has costs and benefits. By itself, it's not necessarily a bad thing. It lets us sleep at night. Without it, we would prolong the awful pangs of embarrassment. We would torture ourselves with regret over the road not taken or over how badly we navigated the road we did take. We would agonize in the aftermath of almost every decision: Did we do the right thing, marry the right person, buy the right house, choose the best car, enter the right career? Yet mindless self-justification, like quicksand, can draw us deeper into disaster. It blocks our ability to even see our errors, let alone correct them. It distorts reality, keeping us from getting all the information we need and assessing issues clearly. It prolongs and widens rifts between lovers, friends, and nations. It keeps us from letting go of unhealthy habits. It permits the guilty to avoid taking responsibility for their deeds. And it keeps many professionals from changing outdated attitudes and procedures that can harm the public.

None of us can avoid making blunders. But we do have the ability to say, "This is not working out here. This is not making sense." To err is human, but humans then have a choice between covering up and fessing up. The choice we make is crucial to what we do next. We are forever being told that we should learn from our mistakes, but how can we learn unless we first admit that we made those mistakes? To do that, we have to recognize the siren song of self-justification. In the next chapter, we will discuss cognitive dissonance, the hardwired psychological mechanism that creates self-justification and protects our certainties, self-esteem, and tribal affiliations. In the chapters that follow, we will elaborate on the most harmful consequences of self-justification: how it exacerbates prejudice and corruption, distorts memory, turns professional confidence into arrogance, creates and perpetuates injustice, warps love, and generates feuds and rifts.

The good news is that by understanding how this mechanism works, we can defeat the wiring. Accordingly, in chapter 8, we will step back and see what solutions emerge for individuals and for relationships. And in chapter 9, we will broaden our perspective to consider the great political issue of our time: the dissonance created when loyalty to the party means supporting a dangerous party leader. The way that citizens resolve that dissonance — by choosing party above nation or by making the difficult but courageous and ethical decision to resist that easy path — has immense consequences for their lives and their country. Understanding is the first step toward finding solutions that will lead to change and redemption. That is why we wrote this book.

1

Cognitive Dissonance:
The Engine of Self-Justification

PRESS RELEASE DATE: NOVEMBER 1, 1993

We didn't make a mistake when we wrote in our previous releases that New York would be destroyed on September 4 and October 14, 1993. We didn't make a mistake, not even a teeny eeny one!

PRESS RELEASE DATE: APRIL 4, 1994

All the dates we have given in our past releases are correct dates given by God as contained in Holy Scriptures. Not one of these dates was wrong . . . Ezekiel gives a total of 430 days for the siege of the city . . . [which] brings us exactly to May 2, 1994. By now, all the people have been forewarned. We have done our job . . .

We are the only ones in the entire world guiding the people to their safety, security, and salvation!

We have a 100 percent track record![1]

It's fascinating, and sometimes funny, to read doomsday predictions, but it's even more fascinating to watch what happens to
the reasoning of true believers when the prediction flops and
the world keeps muddling along. Notice that hardly anyone ever
says, "I blew it! I can't believe how stupid I was to believe that
nonsense"? On the contrary, most of the time the doomsayers
become even more deeply convinced of their powers of prediction. The people who believe that the Bible's book of Revelation
or the writings of the sixteenth-century self-proclaimed prophet
Nostradamus have predicted every disaster from the bubonic
plague to 9/11 cling to their convictions, unfazed by the small
problem that these vague and murky predictions were intelligi-
ble only after the events occurred.

More than half a century ago, a young social psychologist
named Leon Festinger and two associates infiltrated a group
of people who believed the world would end on December 21,
1954.[2] They wanted to know what would happen to the group
when (they hoped!) the prophecy failed. The group's leader,
whom the researchers called Marian Keech, promised that the
faithful would be picked up by a flying saucer and elevated to
safety at midnight on December 20. Many of her followers quit
their jobs, gave away their houses, and disbursed their savings in
anticipation of the end. Who needs money in outer space? Others waited in fear or resignation in their homes. (Mrs. Keech's
husband, a nonbeliever, went to bed early and slept soundly
through the night while his wife and her followers prayed in the
living room.) Festinger made his own prediction: The believers
who had not made a strong commitment to the prophecy — who
awaited the end of the world by themselves at home, hoping they

weren't going to die at midnight — would quietly lose their faith in Mrs. Keech. But those who had given away their possessions and waited with other believers for the spaceship, he said, would increase their belief in her mystical abilities. In fact, they would now do whatever they could to get others to join them.

At midnight, with no sign of a spaceship in the yard, the group felt a little nervous. By 2:00 a.m., they were getting seriously worried. At 4:45 a.m., Mrs. Keech had a new vision: The world had been spared, she said, because of the impressive faith of her little band. "And mighty is the word of God," she told her followers, "and by his word have ye been saved — for from the mouth of death have ye been delivered and at no time has there been such a force loosed upon the Earth. Not since the beginning of time upon this Earth has there been such a force of Good and light as now floods this room."

The group's mood shifted from despair to exhilaration. Many of the group members who had not felt the need to proselytize before December 21 began calling the press to report the miracle. Soon they were out on the streets, buttonholing passersby, trying to convert them. Mrs. Keech's prediction had failed, but not Leon Festinger's.

· · ·

The engine that drives self-justification, the energy that produces the need to justify our actions and decisions — especially the wrong ones — is the unpleasant feeling that Festinger called "cognitive dissonance." Cognitive dissonance is a state of tension that occurs when a person holds two cognitions (ideas, attitudes, beliefs, opinions) that are psychologically inconsistent with each other, such as "Smoking is a dumb thing to do because it could kill me" and "I smoke two packs a day." Dissonance produces mental discomfort that ranges from minor pangs to deep anguish; people don't rest easy until they find a way to reduce it. In this example, the most direct way for a smoker to reduce dissonance is by quitting. But if she has tried to quit and failed, now she must reduce dissonance by convincing herself that smoking isn't really so harmful, that smoking is worth the risk because it helps her relax or prevents her from gaining weight (after all, obesity is a health risk too), and so on. Most smokers manage to reduce dissonance in many such ingenious, if self-deluding, ways.[3]

Dissonance is disquieting because to hold two ideas that contradict each other is to flirt with absurdity, and, as Albert Camus observed, we are creatures who spend our lives trying to convince ourselves that our existence is not absurd. At the heart of it, Festinger's theory is about how people strive to make sense out of contradictory ideas and lead lives that are, at least in their own minds, consistent and meaningful. The theory inspired more than three thousand experiments that, taken together, have transformed psychologists' understanding of how the human mind works. Cognitive dissonance even escaped academia and entered popular culture. The term is everywhere. The two of us have encountered it in political columns, health news stories, magazine articles, a *Non Sequitur* cartoon by Wiley Miller ("Showdown at the Cognitive Dissonance Bridge"), bumper stickers, a TV soap opera, *Jeopardy!,* and a humor column in the *New Yorker* ("Cognitive Dissonances I'm Comfortable With"). Although the expression has been thrown around a lot, few people fully understand its meaning or appreciate its enormous motivational power.

In 1956, one of us (Elliot) arrived at Stanford University as a graduate student in psychology. Festinger had started there that same year as a young professor, and they immediately began working together, designing experiments to test and expand dissonance theory.[4] Their thinking challenged many notions that had been gospel in psychology and among the general public, such as the behaviorist's view that people do things primarily for the rewards they bring, the economist's view that, as a rule, human beings make rational decisions, and the psychoanalyst's view that acting aggressively gets rid of aggressive impulses.

Consider how dissonance theory challenged behaviorism. At the time, most scientific psychologists were convinced that people's actions were governed by reward and punishment. It is certainly true that if you feed a rat at the end of a maze, he will learn the maze faster than if you don't feed him, and if you give your dog a biscuit when she gives you her paw, she will learn that trick faster than if you sit around hoping she will do it on her own. Conversely, if you punish your pup when you catch her peeing on the carpet, she will soon stop doing it. Behaviorists further argued that anything that was associated with reward would become more attractive — your puppy will like you because you give her biscuits — and anything associated with pain would become noxious and undesirable.

Behavioral laws apply to human beings too, of course; no one would stay in a boring job without pay, and if you give your toddler a cookie to stop him from having a tantrum, you have taught him to have another tantrum when he wants a cookie. But, for better or worse, the human mind is more complex than the brain of a rat or a puppy. A dog may appear contrite for having been caught peeing on the carpet, but she will not try to think up justifications for her misbehavior. Humans think — and because we think, dissonance theory demonstrates, our behavior transcends the effects of rewards and punishments and often contradicts them.

To test this observation, Elliot predicted that if people go through a great deal of pain, discomfort, effort, or embarrassment to get something, they will be happier with that "something" than if it came to them easily. For behaviorists, this was a preposterous prediction. Why would people like anything associated with pain? But for Elliot, the answer was obvious: self-justification. The cognition "I am a sensible, competent person" is dissonant with the cognition "I went through a painful procedure to achieve something" — say, join a group — "that turned out to be boring and worthless." Therefore, a person would distort his or her perceptions of the group in a positive direction, trying to find good things about it and ignoring the downside.

It might seem that the easiest way to test this hypothesis would be to rate a number of college fraternities on the basis of how severe their initiations are, then interview members and ask them how much they like their fraternity brothers. If the members of severe-initiation fraternities like their frat brothers more than do members of mild-initiation fraternities, does this prove that severity produces the liking? It does not. It may be just the reverse. If the members of a fraternity regard themselves as being a highly desirable, elite group, they may require a severe initiation to prevent the riffraff from joining. Only those who are highly attracted to the severe-initiation group to begin with would be willing to go through the initiation to get into it. Those who are not excited by a particular fraternity and just want to be in one, any one, will choose fraternities that require mild initiations.

That was why it was essential to conduct a controlled experiment. The beauty of an experiment is the random assignment of people to conditions. Regardless of a person's degree of interest·in joining the group, each participant would be randomly assigned to either the severe-initiation or the mild-initiation condition. If people who went through a tough time to get into a group later find that group to be more attractive than those who got in with no effort, then we would know that it was the effort that caused liking, not the differences in initial levels of interest.

And so Elliot and his colleague Judson Mills conducted just such an experiment.[5] Stanford students were invited to join a group that would be discussing the psychology of sex, but to qualify for admission, they first had to fulfill an entrance requirement. Some of the students were randomly assigned to a severely embarrassing initiation procedure: they had to recite, out loud to the experimenter, lurid, sexually explicit passages from *Lady Chatterley's Lover* and other racy novels. (For conventional 1950s students, this was a painfully embarrassing thing to do.) Others were randomly assigned to a mildly embarrassing initiation procedure: reading aloud sexual words from the dictionary.

After the initiation, each of the students listened to an identical tape recording of a discussion allegedly being held by the group of people they had just joined. Actually, the audiotape was prepared in advance so that the discussion was as boring and worthless as it could be. The discussants talked haltingly, with long pauses, about the secondary sex characteristics of birds — changes in plumage during courtship, that sort of thing. The taped discussants hemmed and hawed, frequently interrupted one another, and left sentences unfinished.

Finally, the students rated the discussion on a number of dimensions. Those who had undergone only a mild initiation saw the discussion for what it was, worthless and dull, and they correctly rated the group members as being unappealing and boring. One guy on the tape, stammering and muttering, admitted that he hadn't done the required reading on the courtship practices of some rare bird, and the mild-initiation listeners were annoyed by him. What an irresponsible idiot! He didn't even do the basic reading! He let the group down! Who'd want to be in a group with him? But those who had gone through a severe initiation rated the discussion as interesting and exciting and the group members as attractive and sharp. They forgave the irresponsible idiot. His candor was refreshing! Who wouldn't want to be in a group with such an honest guy? It was hard to believe that they were listening to the same tape recording. Such is the power of dissonance.

This experiment has been replicated several times by other scientists with a variety of initiation techniques, from electric shock to excessive physical exertion.[6] The results are always the same: severe initiations increase a member's liking for the group. A stunning example of the justification of effort in real life came from an observational study done in the multicultural nation of Mauritius.[7] The annual Hindu festival of Thaipusam includes two rituals: a low-ordeal ritual involving singing and collective prayer, and a severe-ordeal ritual called *kavadi*. "Severe" is something of an understatement. Participants are pierced with needles and skewers, carry heavy bundles, and drag carts that are attached by hooks to their skin for more than four hours. Then they climb a mountain barefoot to reach the temple of Murugan. Afterward, researchers gave both the low-ordeal and severe-ordeal participants the opportunity to anonymously donate money to the temple. The severe-ordeal ritual produced much higher donations than the low-ordeal ritual. The greater the men's pain, the greater their commitment to the temple.

These findings do not mean that people enjoy painful experiences or that they enjoy things because they are associated with pain. What they mean is that if a person voluntarily goes through a difficult or painful experience *in order to attain* some goal or object, that goal or object becomes more attractive. If, on your way to join a discussion group, a flowerpot fell from the open window of an apartment building and hit you on the head, you would not like that discussion group any better. But if you volunteered to get hit on the head by a flowerpot to become a member of the group, you would definitely like the group more.

Believing Is Seeing

> I will look at any additional evidence to confirm the opinion to which I have already come.
> — *Lord Molson, twentieth-century British politician*

Dissonance theory exploded the self-flattering idea that we humans, being *Homo sapiens,* process information logically. On

the contrary; if new information is consonant with our beliefs, we think it is well founded and useful — "Just what I always said!" But if the new information is dissonant, then we
consider it biased or foolish — "What a dumb argument!"
So powerful is the need for consonance that when people are
forced to look at disconfirming evidence, they will find a way
to criticize, distort, or dismiss it so that they can maintain or
even strengthen their existing belief. This mental contortion is
called the "confirmation bias."[8]

Once you are aware of this bias, you'll see it everywhere, including in yourself. Imagine that you are a world-class violin-
ist and your proudest possession is your multimillion-dollar,
three-hundred-year-old Stradivarius. What a thing of beauty it
is! The aged warmth of its tone! Its resonance! Its ease of playability! Now some idiot researcher tries to convince you that
modern violins, some of which cost a mere hundred thousand
dollars or so, are in many ways better than your beloved Strad.
It's such a preposterous claim that you laugh out loud. "Wait,"
the researcher says. "We set up blind tests in hotel rooms with
twenty-one professional violinists and had them wear goggles
that prevented them from knowing whether they were playing a
modern instrument or a Stradivarius, and thirteen of them chose
the new violin as their favorite. Everyone's *least* favorite of the six
instruments tested was a Strad." "Impossible!" you say. "The testing conditions were unrealistic — who can judge a violin's sound
in a hotel room?" So the researcher and her colleagues fine-tune
the study (so to speak). This time they use six three-hundred-
year-old Italian violins and six contemporary ones. They have

ten professional soloists blind-test them for seventy-five minutes in a rehearsal room and then for another seventy-five minutes in a concert hall. The soloists rated the modern violins higher on playability, articulation, and projection, and their guesses as to whether they were playing an old instrument or a new one were no better than chance.[9]

A subsequent study found that listeners, too, prefer the sound of new violins over the allegedly better acoustics of Strads.[10] Strads were rated as sounding better than modern instruments only when the listeners knew what they were hearing. "If you know it's a Strad, you will hear it differently," said the lead researcher. "And you can't turn off that effect."

Will these studies persuade most professional violinists that Strads might be inferior in certain ways to some modern violins? Chances are that the professional violinists will scrutinize the research, looking for flaws. "It's not just the instrument, it's the player," said the concertmaster of the Milwaukee Symphony, whose own Strad is worth five million dollars. "If you're comfortable with an instrument, automatically it's a plus, and the newer instruments, they respond easily. I don't know any great soloist who has a Strad or Guarneri who is trading it in for a new instrument." Not even for a profit of $4,900,000!

The confirmation bias is especially glaring in matters of political observation; we see only the positive attributes of our side and the negative attributes of theirs. Lenny Bruce, the legendary American humorist and social commentator, described this mechanism vividly as he watched the famous 1960 confrontation between Richard Nixon and John Kennedy in the nation's first televised presidential debate:

> I would be with a bunch of Kennedy fans watching the debate and their comment would be, "He's really slaughtering Nixon." Then we would all go to another apartment, and the Nixon fans would say, "How do you like the shellacking he gave Kennedy?" And then I realized that each group loved their candidate so that a guy would have to be this blatant — he would have to look into the camera and say: "I am a thief, a crook, do you hear me, I am the worst choice you could ever make for the Presidency!" And even then his following would say, "Now there's an honest man for you. It takes a big guy to admit that. There's the kind of guy we need for President."[11]

In 2003, after it had become abundantly clear that there were no weapons of mass destruction in Iraq, Democrats and Republicans who had favored going to war (before it began) were thrown into dissonance: We believed the president when he told us Saddam Hussein had WMDs, and we (and he) were wrong. How to resolve this? The majority of Republicans resolved it by refusing to accept the evidence, telling a Knowledge Networks poll that they believed the weapons *had* been found. The survey's baffled director said, "For some Americans, their desire to support the war may be leading them to screen out information that weapons of mass destruction have not been found. Given the intensive news coverage and high levels of public attention to the topic, this level of misinformation suggests that some Americans may be avoiding having an experience of cognitive disso-nance." You bet.[12] Indeed, to this day we occasionally get a query from a reader trying to persuade us that WMDs *were* found. We reply that Bush's top officials — including Donald Rumsfeld, Condoleezza Rice, and Colin Powell — have all acknowledged that there were no WMDs other than a cache of mostly decaying chemical weapons, nothing that warranted going to war over. In his 2010 memoir *Decision Points,* Bush himself wrote, "No one was more shocked and angry than I was when we didn't find the weapons. I had a sickening feeling every time I thought about it. I still do." That "sickening feeling" is cognitive dissonance.

Democrats who had backed President Bush were reducing dissonance too, but in a different way: by actually forgetting that they originally were in favor of the war. Before the inva-sion, about 46 percent of Democrats supported the invasion; by 2006, only 21 percent remembered having done so. Just be-fore the war, 72 percent of Democrats said they thought Iraq had WMDs, but later, only 26 percent remembered having be-lieved this. To maintain consonance, they were saying, in effect, "I knew all along that Bush was lying to us."[13]

Neuroscientists have shown that these biases in thinking are built into the way brains process information — all brains, re-gardless of their owners' political affiliations. In one study, peo-ple were monitored by magnetic resonance imaging (MRI) as they tried to process either dissonant or consonant information about George Bush or John Kerry. Drew Westen and his col-leagues found that the reasoning areas of the brain virtually shut down when participants were confronted with dissonant infor-mation, and the emotion circuits of the brain were activated when consonance was restored.[14] These mechanisms provide a neurological basis for the observation that once our minds are made up, it can be a major effort to change them.

Indeed, even reading information that goes against your point of view can make you all the more convinced you are right. In one experiment, researchers selected people who either favored or opposed capital punishment and asked them to read two scholarly, well-documented articles on the emotionally charged issue of whether the death penalty deters violent crimes. One article concluded that it did, the other that it didn't. If the readers were processing information rationally, they would realize that the issue was more complex than they had previously thought and would therefore move a bit closer to each other in their beliefs about capital punishment as a deterrent. But dissonance theory predicts that the readers would find a way to distort the two articles. They would find reasons to clasp the confirming article to their bosoms and hail it as a highly competent piece of work. And they would be supercritical of the disconfirming article, finding minor flaws and magnifying them into major reasons why they need not be influenced by it. This is precisely what happened. Not only did each side try to discredit the other's arguments; each side became even more committed to its own.[15]

This frequently replicated finding explains why it is so difficult for scientists and health experts to persuade people who are ideologically or politically committed to a belief — such as "climate change is a hoax" — to change their minds even when overwhelming evidence dictates that they should. People who receive disconfirming or otherwise unwelcome information often do not simply resist it; they may come to support their original (wrong) opinion even more strongly—a backfire effect. Once we are invested in a belief and have justified its wisdom, changing our minds is literally hard work. It's much easier to slot that new evidence into an existing framework and do the mental justification to keep it there than it is to change the framework.[16]

The confirmation bias even sees to it that no evidence—the absence of evidence—is evidence for what we believe. When the FBI and other investigators failed to find any evidence whatsoever that the nation had been infiltrated by satanic cults that were ritually slaughtering babies, believers in these cults were unfazed. The absence of evidence, they said, was confirmation of how clever and evil the cult leaders were; they were eating those babies, bones and all.

It's not just fringe cultists and proponents of pop psychology who fall prey to this reasoning. When Franklin D. Roosevelt made the terrible decision to uproot thousands of Japanese Americans and put them in internment camps for the duration of World War II, he did so entirely on the basis of rumors that Japanese Americans were planning to sabotage the war effort. There was no proof then or later to support this rumor. Indeed, the U.S. Army's West Coast commander, General John DeWitt, admitted that the military had no evidence of sabotage or treason against a single Japanese American citizen. Still: "The very fact that no sabotage has taken place," he said, "is a disturbing and confirming indication that such action *will* be taken."[17]

Ingrid's Choice, Nick's Mercedes, and Elliot's Canoe

Dissonance theory came to explain far more than the reasonable notion that people are unreasonable at processing information. It also showed why they continue to be biased after making important decisions.[18] In his illuminating book *Stumbling on Happiness,* social psychologist Dan Gilbert asks us to consider what would have happened at the end of *Casablanca* if Ingrid Bergman had not patriotically rejoined her Nazi-fighting husband but instead remained with Humphrey Bogart in Morocco. Would she, as Bogart tells her in a heart-wrenching speech, have regretted it — "Maybe not today, maybe not tomorrow, but soon, and for the rest of your life"? Or did she forever regret leaving Bogart? Gilbert gathered a wealth of data that shows that the answer to both questions is no, that either decision would have made her happy in the long run. Bogart was eloquent but wrong, and dissonance theory tells us why: Ingrid would have found reasons to justify either choice, along with reasons to be glad she did not make the other.

Once we make a decision, we have all kinds of tools at our disposal to bolster it. When our frugal, unflashy friend Nick traded in his eight-year-old Honda Civic on a sudden impulse and bought a new, fully loaded Mercedes, he began behaving oddly (for Nick). He started criticizing his friends' cars, saying things like "Isn't it about time you traded in that wreck? Don't you think you deserve the pleasure of driving a well-engineered machine?" and "You know, it's really unsafe to drive little cars. If you got in an accident, you could be killed. Isn't your life worth an extra few thousand dollars? You have no idea how much peace of mind it brings me to know that my family is safe because I'm driving a solid automobile."

It's possible that Nick simply got bitten by the safety bug and decided, coolly and rationally, that it would be wonderful if all his friends drove a great car like the Mercedes. But we don't think so. His behavior was so uncharacteristic that we suspected that he was reducing the dissonance he felt over impulsively spending a big chunk of his life's savings on what he would once have referred to as "just a car." In addition, he did this just when his kids were about to go to college, an event that would put a strain on his bank account. So Nick began marshaling arguments to justify his decision: "The Mercedes is a wonderful machine; I've worked hard all my life and I deserve it; besides, it's so safe." And if he could persuade his cheapskate friends to buy one too, he would feel doubly justified. Like Mrs. Keech's converts, he began to proselytize.

Nick's need to reduce dissonance was increased by the irrevocability of his decision; he could not unmake that decision without losing a lot of money. Some scientific evidence for the power of irrevocability comes from a clever study of the mental maneuverings of gamblers at a racetrack. The racetrack is an ideal place to study irrevocability because once you've placed your bet, you can't go back and tell the nice man behind the window you've changed your mind. In this study, the researchers simply intercepted people who were standing in line to place two-dollar bets and other people who had just left the window. The investigators asked them how certain they were that their horses would win. The bettors who had placed their bets were far more certain about their choice than the folks waiting in line.[19] Yet nothing had changed except the finality of placing the bet. People become more certain they are right about something they just did if they can't undo it.

You can see one immediate benefit of understanding how dissonance works: Don't listen to Nick. The more costly a decision in terms of time, money, effort, or inconvenience and the more irrevocable its consequences, the greater the dissonance and the greater the need to reduce it by overemphasizing the good things about the choice made. Therefore, when you are about to make a big purchase or an important decision — which car or computer to buy, whether to undergo plastic surgery, or whether to sign up for a costly self-help program — don't ask someone who has just done it. That person will be highly motivated to convince you that it is the right thing to do. Ask people who have spent twelve years and fifty thousand dollars on a particular therapy if it helped, and most will say, "Dr. Weltschmerz is wonderful! I would *never* have [found true love] [got a new job] [taken up tap dancing] if it hadn't been for him." After investing all that time and money, they aren't likely to say, "Yeah, I saw Dr. Weltschmerz for twelve years, and boy, was it ever a waste." Behavioral economists have shown how reluctant people are to accept these *sunk costs* — investments of time or money that they've sunk into an experience or relationship. Rather than cutting their losses, most people will throw good money after bad in hopes of recouping those losses and justifying their original decision. Therefore, if you want advice on what product to buy, ask someone who is still gathering information and is still open-minded. And if you want to know whether a program will help you, don't rely on testimonials; get the data from controlled experiments.

Self-justification is complicated enough when it follows our conscious choices and we know we can expect it. But it also occurs in the aftermath of things we do for unconscious reasons, when we haven't a clue about why we hold some belief or cling to some custom but are too proud to admit it. In the introduction, we described the custom of the Dinka and Nuer tribes of the Sudan, who extract several of the permanent front teeth of their children — a painful procedure, done with a fishhook. Anthropologists suggest that this tradition originated during an epidemic of lockjaw; missing front teeth would enable sufferers to get some nourishment. But if that was the reason, why in the world would the villagers continue this custom once the danger had passed?

A practice that makes no sense at all to outsiders makes perfect sense when seen through the lens of dissonance theory. During the epidemic, the villagers might have begun extracting the front teeth of all their children so that if any of them later contracted tetanus, the adults would be able to feed them. But this is a painful thing to do to children, and in any case, only some would become infected. To further justify their actions, to themselves and their children, the villagers needed to bolster the decision by adding benefits to the procedure after the fact. Thus, they might convince themselves that missing teeth had aesthetic value — "Say, that sunken-chin look is really quite attractive" — and they might even turn the surgical ordeal into a rite of passage into adulthood. And, indeed, that is just what happened. "The toothless look is beautiful," the villagers say. "People who have all their teeth are ugly; they look like cannibals who would eat a person. A full set of teeth makes a man look like a donkey." The toothless look has other aesthetic advantages: "We like the hissing sound it creates when we speak." And adults console frightened children by saying, "This ritual is a sign of maturity."[20] The original medical justification for the practice is long gone. The psychological self-justification remains.

People want to believe that, being smart and rational individuals, they know why they make the choices they do, so they are not always happy when you tell them the actual reason for their actions. Elliot learned this firsthand after that initiation experiment. "After each participant had finished," he recalls, "I explained the study in detail and went over the theory carefully. Although everyone who went through the severe initiation said that they found the hypothesis intriguing and that they could see how most people would be affected in the way I predicted, they all took pains to assure me that their preference for the group had nothing to do with the severity of the initiation. They each claimed that they liked the group because that's the way they really felt. Yet almost all of them liked the group more than any of the people in the mild-initiation condition did."

No one is immune to the need to reduce dissonance, even those who know the theory inside out. Elliot tells this story: "When I was a young professor at the University of Minnesota, my wife and I tired of renting apartments. So, in December, we set out to buy our first home. We could find only two reasonable houses in our price range. One was older, charming, and within walking distance of the campus. I liked it a lot, primarily because it meant that I could have my students over for research meetings, serve beer, and play the role of the hip professor. But that house was in an industrial area, without a lot of space for our children to play. The other choice was a tract house, newer but totally without distinction. It was in the suburbs, a thirty-minute drive from campus but only a mile from a lake. After going back and forth on that decision for a few weeks, we decided on the house in the suburbs.

"Shortly after moving in, I noticed an ad in the newspaper for a used canoe and immediately bought it as a surprise for my wife and kids. When I drove home on a freezing, bleak January day with the canoe lashed to the roof of my car, my wife took one look and burst into laughter. 'What's so funny?' I asked. She said, 'Ask Leon Festinger!' Of course! I had felt so much dissonance about buying the house in the suburbs that I needed to do something right away to justify that purchase. I somehow managed to forget that it was the middle of winter and that, in Minneapolis, it would be months before the frozen lake would thaw out enough for the canoe to be usable. But, in a sense, without my quite realizing it, I used that canoe anyway. All winter, even as it sat in the garage, its presence made me feel better about our decision."

Spirals of Violence — and Virtue

Feeling stressed? One internet source teaches you how to make your own little Dammit Doll, which "can be thrown, jabbed, stomped and even strangled till all the frustration leaves you." A little poem goes with it:

> Whenever things don't go so well,
> And you want to hit the wall and yell,
> Here's a little dammit doll that you can't do without.
> Just grasp it firmly by the legs and find a place to slam it.
> And as you whack the stuffing out, yell, "Dammit, dammit,
>   dammit!"

The Dammit Doll reflects one of the most entrenched convictions in our culture, fostered by the psychoanalytic belief in the benefits of catharsis: expressing anger or behaving aggressively gets rid of anger. Throw that doll, hit a punching bag, shout at your spouse; you'll feel better afterward. Actually, decades of experimental research have found exactly the opposite: when people vent their feelings aggressively, they often feel worse, pump up their blood pressure, and make themselves even angrier.[21]

Venting is especially likely to backfire when a person commits an aggressive act against another person directly, and that is exactly what cognitive dissonance theory would predict. When you do anything that harms others — get them in trouble, verbally abuse them, or punch them out — a powerful new factor comes into play: the need to justify what you did. Take a boy who goes along with a group of his fellow seventh graders who are taunting and bullying a weaker kid who did them no harm. The boy likes being part of the gang but his heart really isn't in the bullying. Later, he feels some dissonance about what he did. "How can a decent kid like me," he wonders, "have done such a cruel thing to a nice, innocent little kid like him?" To reduce dissonance, he will try to convince himself that the victim is neither nice nor innocent: "He is such a nerd and a crybaby. Besides, he would have done the same to me if he had the chance." Once the boy starts down the path of blaming the victim, he becomes more likely to beat up on the victim with even greater ferocity the next chance he gets. Justifying his first hurtful act sets the stage for more aggression. That's why the catharsis hypothesis is wrong.

The results of the first experiment that demonstrated this actually came as a complete surprise to the investigator. Michael Kahn, then a graduate student in clinical psychology at Harvard, designed an ingenious experiment that he was sure would demonstrate the benefits of catharsis. Posing as a medical technician, Kahn took polygraph and blood pressure measurements from college students, one at a time, allegedly as part of a medical experiment. As he was taking these measurements, Kahn feigned annoyance and made some insulting remarks to the students (having to do with their mothers). The students got angry; their blood pressure soared. In the experimental condition, the students were allowed to vent their anger by informing Kahn's supervisor of his insults; thus, they believed they were getting him in big trouble. In the control condition, the students did not get a chance to express their anger.

Kahn, a good Freudian, was astonished by the results: Catharsis was a total flop in terms of making people feel better. The people who were allowed to express their anger about Kahn felt far greater animosity toward him than those who were not given that opportunity. In addition, although everyone's blood pressure went up during the experiment, subjects who expressed their anger showed even greater elevations; the blood pressure of those who were not allowed to express their anger soon returned to normal.[22] Seeking an explanation for this unexpected pattern, Kahn discovered dissonance theory, which was just getting attention at the time, and realized it could beautifully account for his results. Because the students thought they had gotten the technician in serious trouble, they had to justify their actions by convincing themselves that he deserved it, thus increasing their anger — and their blood pressure.

Children learn to justify their aggressive actions early; a child hits his younger sibling, who starts to cry, and immediately the boy claims, "But he started it! He deserved it!" Most parents find these childish self-justifications to be of no great consequence, and usually they aren't. But it is sobering to realize that the same mechanism underlies the behavior of gangs who bully weaker children, employers who mistreat workers, lovers who abuse each other, police officers who continue beating a suspect who has surrendered, tyrants who imprison and oppress ethnic minorities, and soldiers who commit atrocities against civilians. In all these cases, a vicious circle is created: Aggression begets self-justification, which begets more aggression. Fyodor Dostoyevsky understood perfectly how this process works. In *The Brothers Karamazov,* he has Fyodor Pavlovitch, the brothers' scoundrel of a father, recall "how he had once in the past been asked, 'Why do you hate so-and-so so much?' And he had answered them, with his shameless impudence, 'I'll tell you. He has done me no harm. But I played him a dirty trick, and ever since I have hated him.'"

Fortunately, dissonance theory also shows us how a person's generous actions can create a spiral of benevolence and compassion, a "virtuous circle." When people do a good deed, particularly when they do it on a whim or by chance, they will come to see the beneficiary of their generosity in a warmer light. Their cognition that they went out of their way to do a favor for this person is dissonant with any negative feelings they might have had about him. In effect, after doing the favor, they ask themselves: "Why would I do something nice for a jerk? Therefore, he's not as big a jerk as I thought he was — as a matter of fact, he is a pretty decent guy who deserves a break."

Several experiments have supported this prediction. In one, college students participated in a contest in which they won substantial sums of money. Afterward, the experimenter approached one-third of them and explained that he was using his own funds for the experiment and was running short, which meant he might be forced to close down the experiment prematurely. He asked, "As a special favor to me, would you mind returning the money you won?" (They all agreed.) A second group was also asked to return the money, but this time it was the departmental secretary who made the request, explaining that the psychology department's research fund was running low. (They still all agreed.) The remaining participants were not asked to return their winnings at all. Finally, everyone filled out a questionnaire that included an opportunity to rate the experimenter. Participants who had been cajoled into doing a special favor for him liked him the best; they convinced themselves he was a particularly fine, deserving fellow. The others thought he was pretty nice but not anywhere near as wonderful as the people who had done him a personal favor believed.[23]

The mechanism of the virtuous circle starts early. In a study of four-year-olds, children were given one sticker each and then introduced to a doggie puppet "who is sad today"; some of the children were told they had to give the sticker to Doggie, while others had a choice of whether or not to give the sticker away. Later, the children were given three stickers each, introduced to another sad puppet, Ellie, and told they could share up to three stickers with her. The children who had been *allowed to choose* to be generous to the sad doggie shared more with Ellie than the children who had been *instructed* to share. In other words, once children saw themselves as generous kids, they continued to behave generously.[24]

Although scientific research on the virtuous circle is relatively new, the general idea may have been discovered in the eighteenth century by Benjamin Franklin, a serious student of human nature as well as science and politics. While serving in the Pennsylvania legislature, Franklin was disturbed by the opposition and animosity of a fellow legislator. So he set out to win him over. He didn't do it, he wrote, by "paying any servile respect to him" — that is, by doing the other man a favor — but by inducing his target to do a favor for *him*. He asked the man to loan him a rare book from his library.

> He sent it immediately and I returned it in about a week with another note, expressing strongly my sense of the favor. When we next met in the House, he spoke to me (which he had never done before), and with great civility; and he ever after manifested a readiness to serve me on all occasions, so that we became great friends, and our friendship continued to his death. This is another in-

stance of the truth of an old maxim I had learned, which says, "He that has once done you a kindness will be more ready to do you another than he whom you yourself have obliged."[25]

• • •

Dissonance is bothersome under any circumstances, but it is most painful to people when an important element of their self-concept is threatened — typically when they do something that is inconsistent with their view of themselves.[26] If a celebrity you admire is accused of an immoral act, you will feel a pang of dissonance, and the more you liked and admired that person, the greater the dissonance you'll feel. (Later in this book, we'll discuss the massive dissonance felt by Michael Jackson's many fans upon hearing compelling evidence of his sexual relationships with young boys.) But that's nothing compared to how you would feel if *you* did the immoral thing. If you regard yourself as a person of high integrity and you do something that harms another person, you'll feel a much more devastating rush of dissonance than you would on hearing about a favorite movie star's transgression. After all, you can always abandon your allegiance to a celebrity or find another hero. But if you violate your own values, you'll feel much greater dissonance because, at the end of the day, you have to go on living with yourself.

In a sweet demonstration of how the need for self-esteem trumps the virtue of realistic modesty, the great majority of people think they are "better than average" — we might call this the Lake Wobegone effect. They say they are better than average in all kinds of ways — smarter, nicer, more ethical, funnier, more competent, more humble, even better drivers.[27] Their efforts at reducing dissonance are therefore designed to preserve these positive self-images.[28] When Mrs. Keech's doomsday predictions failed, imagine the excruciating dissonance her committed followers felt — "I am a smart person" clashed with "I just did an incredibly stupid thing: I gave away my house and possessions and quit my job because I believed a crazy woman." To reduce that dissonance, her followers could either modify their opinion of their intelligence or justify the incredibly stupid thing they had just done. It's not a close contest; justification wins by three lengths. Mrs. Keech's true believers saved their self-esteem by deciding they hadn't done anything stupid; actually, they had been really smart to join this group because their faith saved the world from destruction. In fact, if others were smart, they would join too. Where's that busy street corner?

None of us is off the hook on this one. *We* might feel amused at *them,* those foolish people who believe fervently in doomsday predictions, but, as political scientist Philip Tetlock shows in his book *Expert Political Judgment,* even professionals who are in the business of economic and political forecasting are usually no more accurate than us untrained folks — or Mrs. Keech, for that matter.[29]

And what do these experts do when *their* prophecies are disconfirmed? In 2010, a coalition of twenty-three prominent economists, fund managers, academics, and journalists signed a letter opposing the Federal Reserve's practice of buying long-term debt as a way of pushing down long-term interest rates. This practice risks "currency debasement and inflation" and fails to create jobs, the experts stated, and therefore should be "reconsidered and discontinued." Four years later, inflation was still low (indeed lower than the Federal Reserve's goal of 2 percent), unemployment had fallen sharply, job growth was improving, and the stock market was soaring. Accordingly, reporters went back to the letter's signers and asked, Have you changed your minds? Of the twenty-three signatories, fourteen didn't reply. The other nine said their views were unchanged; they were just as worried about inflation now as they had ever been. Like the failed doomsday prophets, they had clever self-justifications for not admitting they had been wrong, very wrong. One said the nation *had* had inflation; it just hadn't shown up yet in consumer prices. One, using what he later admitted were bogus statistics, claimed the country was in the midst of double-digit inflation. One said that "official numbers err" and that inflation was really much higher than the Bureau of Labor Statistics claimed. And several, with echoes of doomsday, said their *prediction* was right but the *date* was wrong: "High inflation will come someday; we just haven't said when."[30]

Experts can sound pretty impressive, especially when they bolster their claims by citing their years of training and experience in a field. Yet hundreds of studies have shown that, compared to predictions based on actuarial data, predictions based on an expert's years of training and personal experience are rarely better than chance. But when an expert is wrong, the centerpiece of his or her professional identity is threatened. Therefore, dissonance theory predicts that the more self-confident and famous experts are, the less likely they will be to admit mistakes. And that is just what Tetlock found. Experts reduced the dissonance caused by their failed forecasts by coming up with expla-

nations of why they would have been right "if only" — if only
that improbable calamity had not intervened; if only the timing
of events had been different; if only blah-blah-blah.

Dissonance reduction operates like the burner on a stove,
keeping our self-esteem bubbling along. That is why we are usually oblivious to the self-justifications, the little lies to ourselves
that prevent us from even acknowledging that we made mistakes
or foolish decisions. But dissonance theory applies to people
with low self-esteem too, to people who consider themselves to
be schnooks, crooks, or incompetents. They are not surprised
when their behavior confirms their negative self-image. When
they make wrong-headed predictions or go through severe initiations to get into what turns out to be dull groups, they merely
say, "Yup, I screwed up again; that's just like me." A used-car
dealer who knows that he is dishonest does not feel dissonance
when he conceals the dismal repair record of the car he is trying
to unload; a woman who believes she is unlovable does not feel
dissonance when a man rejects her; a con man does not experience dissonance when he cheats his grandmother out of her life
savings.

Our convictions about who we are carry us through the day,
and we are constantly interpreting the things that happen to us
through the filter of those core beliefs. When those beliefs are violated, even by a good experience, it causes us discomfort. An appreciation of the power of self-justification helps us understand
why people who have low self-esteem or who simply believe that
they are incompetent in some domain are not totally overjoyed
when they do something well; on the contrary, they often feel
like frauds. If the woman who believes she is unlovable meets a

terrific guy who starts pursuing her seriously, she will feel momentarily pleased, but that pleasure is likely to be tarnished by a rush of dissonance: "What does he see in me?" Her resolution is unlikely to be "How nice; I must be more appealing than I thought I was." More likely, it will be "As soon as he discovers the real me, he'll dump me." She will pay a high psychological price to have that consonance restored.

Indeed, several experiments find that most people who have low self-esteem or a low estimate of their abilities do feel uncomfortable with dissonant successes and dismiss them as accidents or anomalies.[31] This is why they seem so stubborn to friends and family members who try to cheer them up. "Look, you just won the Pulitzer Prize in literature! Doesn't *that* mean you're good?" "Yeah, it's nice, but just a fluke. I'll never be able to write another word, you'll see." Self-justification, therefore, will protect high self-esteem to avoid dissonance, but it will also protect low self-esteem if that is a default self-perception.

The Pyramid of Choice

Imagine two young men who are identical in terms of attitudes, abilities, and psychological health. They are reasonably honest and have the same middling attitude toward, say, cheating—they think it is not a good thing to do, but there are worse crimes in the world. Now they are both in the midst of taking an exam that will determine whether they will get into graduate school. They each draw a blank on a crucial essay question. Failure looms . . . at which point each one gets an easy opportunity to cheat by reading another student's answers. The two young men struggle with temptation. After a long moment of anguish, one yields, and the other resists. Their decisions are a hairsbreadth apart; it could easily have gone the other way for each of them. Each gains something important, but at a cost: One gives up integrity for a good grade; the other gives up a good grade to preserve his integrity.

Now the question is: How will they feel about cheating one week later? Each student has had ample time to justify the course of action he took. The one who yielded to temptation will decide that cheating is not so great a crime. He will say to himself: "Hey, everyone cheats. It's no big deal. And I really needed to do this for my future career." But the one who resisted temptation will decide that cheating is far more immoral than he originally thought. "In fact," he'll tell himself, "people who cheat are disgraceful. In fact, people who cheat should be permanently expelled from school. We have to make an example of them."

By the time the students are through with their increasingly intense levels of self-justification, two things have happened. One, they are now a great distance apart from each other, and two, they have internalized their beliefs and are convinced that they have always felt that way.[32] It is as if they started off at the top of a pyramid a millimeter apart, but by the time they have finished justifying their individual actions, they have slid to the bottom and now stand at opposite corners of its base. The one who didn't cheat considers the other to be totally immoral, and the one who cheated thinks the other is hopelessly puritanical. This process illustrates how people who have been sorely tempted, battled temptation, and almost given in to it — but resisted at the eleventh hour — come to dislike, even despise, those who did not succeed in the same effort. It's the people who *almost* decide to live in glass houses who throw the first stones.

When a cheating scandal occurred at the high-achieving, high-pressure Stuyvesant High School in New York City — seventy-one students were caught exchanging exam answers — students gave a *New York Times* reporter a litany of self-justifications that allowed them to keep seeing themselves as smart students of integrity: "It's like, 'I'll keep my integrity and fail this test,'" said one. "No. No one wants to fail a test. You could study for two hours and get an 80, or you could take a risk and get a 90." He redefined cheating as "taking a risk." For others, cheating was a "necessary evil." For many, it was "helping classmates in need." When one girl finally realized her classmates had been relying on her to write their papers for them, she said, "I respect them and think they have integrity . . . [but] sometimes the only way you could've gotten there is to kind of botch your ethics for a couple things." *Kind of botch your ethics?* Minimizing ethical violations is a popular form of self-justification. Hana Beshara started a website that pirated films and TV shows for instant free downloading, in clear violation of the copyright laws. Caught, she was sent to prison for sixteen months for conspiracy and criminal copyright infringement. But did she make a mistake or do wrong? No. "I never imagined it going criminal," she told a reporter. "It didn't seem like it was something to be bothered with. Even if it is wrong."[33]

The metaphor of the pyramid applies to most important decisions involving moral choices or life options. Instead of cheating on an exam, you can substitute deciding to begin a casual affair (or not), take steroids to improve your athletic ability (or not), stay in a troubled marriage (or not), lie to protect your employer and job (or not), have children (or not), pursue a demanding career (or stay home with the kids), decide that a sensational allegation against a celebrity you admire is false (or true). When the person at the top of the pyramid is uncertain, when there are benefits and costs for both choices, then he or she will feel a particular urgency to justify the choice made. But by the time the person is at the bottom of the pyramid, ambivalence will have morphed into certainty, and he or she will be miles away from anyone who took a different route.

This process blurs the distinction that people like to draw between "us good guys" and "those bad guys." Often, when standing at the top of the pyramid, we are faced not with a black-or-white, go-or-no-go decision but with gray choices whose consequences are shrouded. The first steps along the path are morally ambiguous, and the right decision is not always clear. We make an early, apparently inconsequential decision, and then we justify it to reduce the ambiguity of the choice. This starts a process of entrapment — action, justification, further action — that increases our intensity and commitment and may end up taking us far from our original intentions or principles.

It certainly worked that way for Jeb Stuart Magruder, Richard Nixon's special assistant. Magruder, a key player in the plot to burglarize the Democratic National Committee headquarters in the Watergate complex, concealed the White House's involvement and lied under oath to protect himself and others responsible. When Magruder was first hired, Nixon's adviser Bob Haldeman did not mention that perjury, cheating, and breaking the law were part of the job description. If he had, Magruder almost certainly would have refused. How, then, did he end up as a central player in the Watergate scandal? In hindsight, it is easy to say he should have known or he should have drawn the line the first time they asked him to do something illegal.

In his autobiography, Magruder describes his initial meeting with Bob Haldeman at San Clemente. Haldeman flattered and charmed him. "Here you're working for something more than just to make money for your company," Haldeman told him. "You're working to solve the problems of the country and the world. Jeb, I sat with the President on the night the first astronauts stepped onto the moon . . . I'm part of history being made." At the end of a day of meetings, Haldeman and Magruder left the compound to go to the president's house. Haldeman was enraged that his golf cart was not right there awaiting him, and he gave his assistant a "brutal chewing out," threatening to fire the guy if he couldn't do his job. Magruder couldn't believe what he was hearing, especially since it was a beautiful evening and a short walk to their destination. At first Magruder thought Haldeman's tirade was rude and excessive. But before long, wanting the job as much as he did, Magruder was justifying Haldeman's behavior: "In just a few hours at San Clemente I had been struck by the sheer *perfection* of life there . . . After you have been spoiled like that for a while, something as minor as a missing golf cart can seem a major affront."[34]

And so, before dinner and even before having been offered a job, Magruder was hooked. It was a tiny first step, but he was on the road to Watergate. Once he was in the White House, he went along with all of the small ethical compromises that just about all politicians justify in the goal of serving their party. Then, when

Magruder and others were working to reelect Nixon, G. Gordon
Liddy entered the picture, hired by attorney general John Mitchell to be Magruder's general counsel. Liddy was a wild card, a
James Bond wannabe. His first plan to ensure Nixon's reelection
was to spend one million dollars to hire "squads" to rough up
demonstrators, kidnap activists who might disrupt the Republican Convention, sabotage the Democratic Convention, hire
"high-class" prostitutes to entice and then blackmail leading
Democrats, and break into Democratic offices and install elec-
tronic-surveillance devices and wiretaps.

Mitchell disapproved of the more extreme aspects of this
plan; further, he said, it was too expensive. So Liddy returned
with a proposal merely to break into the DNC offices at the
Watergate complex and install wiretaps. This time Mitchell approved, and the others went along. How did they justify breaking the law? "If [Liddy] had come to us at the outset and said,
'I have a plan to burglarize and wiretap Larry O'Brien's office,'
we might have rejected the idea out of hand," wrote Magruder.
"Instead, he came to us with his elaborate call girl/kidnapping/
mugging/sabotage/wiretapping scheme, and we began to tone
it down, always with a feeling that we should leave Liddy a little
something — we felt we needed him, and we were reluctant to
send him away with nothing." Finally, Magruder added, Liddy's
plan was approved because of the paranoid climate in the White
House: "Decisions that now seem insane seemed at the time to
be rational . . . We were past the point of halfway measures or
gentlemanly tactics."[35]

When Magruder first entered the White House, he was a decent man. But, one small step at a time, he went along with dis-

honest actions, justifying each one as he did. He was entrapped in pretty much the same way as the three thousand people who took part in the famous experiment created by social psychologist Stanley Milgram.[36] In Milgram's original version, two-thirds of the participants administered what they thought were life-threatening levels of electric shock to another person simply because the experimenter kept saying, "The experiment requires that you continue." This experiment is almost always described as a study of obedience to authority. Indeed it is. But it is more than that; it is also a demonstration of long-term results of self-justification.[37]

Imagine that a distinguished-looking man in a white lab coat walks up to you and offers you twenty dollars to participate in a scientific experiment. He says, "I want you to inflict five hundred volts of incredibly painful shock to another person to help us understand the role of punishment in learning." Chances are you would refuse; the money isn't worth it to harm another person, even for science. A few people would do it for twenty bucks, but most would tell the scientist where he could stick his money.

Now suppose the scientist lures you along more gradually. Suppose he offers you twenty dollars to administer a minuscule amount of shock, say ten volts, to a fellow in the adjoining room to see if this zap will improve the man's ability to learn. The experimenter even tries the ten volts on you, and you can barely feel it. So you agree. It's harmless and the study seems pretty interesting. (Besides, you've always wanted to know whether spanking your kids will get them to shape up.) You go along for the moment, and now the experimenter tells you that if the learner gets the wrong answer, you must move to the next toggle switch, which delivers a shock of twenty volts. Again, it's a small and harmless jolt. Because you just gave the learner ten, you see no reason why you shouldn't give him twenty. And once you give him twenty, you say to yourself, "Thirty isn't much more than twenty, so I'll go to thirty." He makes another mistake, and the scientist says, "Please administer the next level — forty volts."

Where do you draw the line? When do you decide enough is enough? Will you keep going to 450 volts, or even beyond that, to a switch marked XXX DANGER? When people were asked in advance how far they imagined they would go, almost no one said they would go to 450. But when they were actually in the situation, two-thirds of them went all the way to the maximum level they believed was dangerous. They did this by justifying each step as they went along: "This small shock doesn't hurt; twenty isn't much worse than ten; if I've given twenty, why not thirty?" With each justification, they committed themselves further. By the time people were administering what they believed were strong shocks, most found it difficult to justify a decision to quit. Participants who resisted early in the study, questioning the validity of the procedure itself, were less likely to become trapped by it and more likely to walk out.

The Milgram experiment shows us how ordinary people can end up doing immoral and harmful things through a chain reaction of behavior and subsequent self-justification. When we, as observers, look at them in puzzlement or dismay, we fail to realize that we are often looking at the end of a long, slow process down that pyramid. At his sentencing, Magruder said to Judge John Sirica: "I know what I have done, and Your Honor knows what I have done. Somewhere between my ambition and my ideals, I lost my ethical compass." How do you get an honest man to lose his ethical compass? You get him to take one step at a time, and self-justification will do the rest.

• • •

Knowing how dissonance works won't make any of us automatically immune to the allure of self-justification, as Elliot learned when he bought that canoe in a Minnesota January. You can't say to people, as he did after the initiation experiments, "See how you reduced dissonance? Isn't that interesting?" and expect them to reply, "Oh, thank you for showing me the real reason I like the group. That sure makes me feel smart!" To preserve our belief that we are smart, all of us will occasionally do dumb things. We can't help it. We are wired that way.

But this does not mean that we are doomed to keep striving to justify our actions after the fact, to be like Sisyphus, never reaching the top of the hill of self-acceptance. A richer understanding of how and why our minds work as they do is the first step toward breaking the self-justification habit. And that, in turn, requires us to be more mindful of our behavior and the reasons for our choices. It takes time, self-reflection, and willingness.

In 2003, the conservative columnist William Safire wrote that a "psychopolitical challenge" voters often face was "how to deal with cognitive dissonance."[38] He began with a story of his own such challenge. During Bill Clinton's administration, Safire recounted, he had criticized Hillary Clinton for trying to conceal the identity of the members of her health-care task force. He wrote a column castigating her efforts at secrecy, which he said were toxic to democracy. No dissonance there; those bad Democrats are always doing bad things. Six years later, however, he found that he was "afflicted" by cognitive dissonance when Vice President Dick Cheney, a fellow conservative Republican whom Safire admired, insisted on keeping the identity of his energy-policy task force a secret. What did Safire do? Because of his awareness of dissonance and how it works, he took a deep breath, hitched up his trousers, and did the tough but virtuous thing: He wrote a column publicly criticizing Cheney's actions. The irony is that because of his criticism of Cheney, Safire received several laudatory letters from liberals — which, he admitted, produced enormous dissonance. Oh Lord, he'd done something *those* people approved of?

Safire's ability to recognize his own dissonance and resolve it by doing the fair thing is rare. As we will see, his willingness to concede that his own side made a mistake is something that few are prepared to do. Instead, conservatives and liberals alike will bend over backward to reduce dissonance in a way that is favorable to them and their team. The specific tactics vary, but our efforts at self-justification are all designed to serve our need to feel good about what we have done, what we believe, and who we are.

2

Pride and Prejudice . . .
and Other Blind Spots

And why do you look at the speck in your brother's eye,
but do not consider the plank in your own eye?
 — *Matthew 7:3*

When the public learned that Supreme Court justice Antonin
Scalia was flying to Louisiana on a government plane to go duck
hunting with Vice President Dick Cheney despite Cheney's having a pending case before the Supreme Court, there was a flurry
of protest at Scalia's apparent conflict of interest. Scalia himself
was indignant at the suggestion that his ability to assess the con-
stitutionality of Cheney's claim — that the vice president was legally entitled to keep the details of his energy task force secret
 — would be tainted by the ducks and the perks. In a letter to the
Los Angeles Times explaining why he would not recuse himself,
Scalia wrote, "I do not think my impartiality could reasonably
be questioned."

• • •

Neuropsychologist Stanley Berent and neurologist James Albers were hired by CSX Transportation and Dow Chemical to investigate railroad workers' claims that chemical exposure had caused permanent brain damage and other medical problems. More than six hundred railroad workers in fifteen states had been diagnosed with a form of brain damage following heavy exposure to chlorinated hydrocarbon solvents. CSX paid more than $170,000 to Berent and Albers's consulting firm for research that eventually disputed a link between exposure to the company's industrial solvents and brain damage. While conducting their study, which involved reviewing the workers' medical files without the workers' informed consent, the two scientists served as expert witnesses for law firms representing CSX in lawsuits filed by workers. Berent saw nothing improper in his research, which he claimed "yielded important information about solvent exposure." Berent and Albers were subsequently reprimanded by the federal Office of Human Research Protections for their conflict of interest in this case.[1]

•   •   •

When you enter the Museum of Tolerance in Los Angeles, you find yourself in a room of interactive exhibits designed to identify the people you can't tolerate. The familiar targets are there (blacks, women, Jews, gays), but also short people, fat people, blond-female people, disabled people ... You watch a video on the vast variety of prejudices designed to convince you that all human beings have at least a few, and then you are invited to enter the museum proper through one of two doors, one marked PREJUDICED, the other marked UNPREJUDICED. The latter door is locked, in case anyone misses the point, and occasionally some people do. When we were visiting the museum one afternoon, we were treated to the sight of four Hasidic Jews pounding angrily on the Unprejudiced door, demanding to be let in.

• • •

The brain is designed with blind spots, optical and psychological, and one of its cleverest tricks is to confer on its owner the comforting delusion that he or she does not have any. In a sense, dissonance theory is a theory of blind spots — of how and why people unintentionally blind themselves so that they fail to notice vital events and information that might make them question their behavior or their convictions. Along with the confirmation bias, the brain comes packaged with other self-serving habits that allow us to justify our own perceptions and beliefs as being accurate, realistic, and unbiased. Social psychologist Lee Ross named this phenomenon "naive realism," the inescapable conviction that we perceive objects and events clearly, "as they really are."[2] We assume that other reasonable people see things the same way we do. If they disagree with us, they obviously aren't seeing clearly. Naive realism creates a logical labyrinth because it presupposes two things: One, people who are open-minded and fair ought to agree with a reasonable opinion, and, two, any opinion I hold must be reasonable; if it weren't, I wouldn't hold it. Therefore, if I can just get my opponents to sit down here and listen to me explain how things really are, they will agree with me. And if they don't, it must be because they are biased.

Ross knows whereof he speaks from both his laboratory experiments and his efforts to reduce the bitter conflict between Israelis and Palestinians. Even when each side recognizes that the other side perceives the issues differently, each thinks that the other side is biased while they themselves are objective and that their own perceptions of reality should provide the basis for settlement. In one experiment, Ross took peace proposals created by Israeli negotiators, labeled them as Palestinian proposals, and asked Israeli citizens to judge them. "The Israelis liked the Palestinian proposal attributed to Israel more than they liked the Israeli proposal attributed to the Palestinians," he says. "If your own proposal isn't going to be attractive to you when it comes from the other side, what chance is there that the *other* side's proposal is going to be attractive when it actually comes from the other side?"[3] Closer to home, social psychologist Geoffrey Cohen found that Democrats will endorse an extremely restrictive welfare proposal, one usually associated with Republicans, if they think it has been proposed by the Democratic Party, and Republicans will support a generous welfare policy if they think it comes from the Republican Party.[4] Label the same proposal as coming from the other side, and you might as well be asking people to support a policy proposed by Hitler, Stalin, or Attila the Hun. None of the people in Cohen's study were aware of their blind spot — that they were being influenced by their party's position. Instead, they all claimed that their beliefs followed logically from their own careful study of the policy at hand, guided by their general philosophy of government.

It's immensely hard to overcome this blind spot, even when doing so is part of your job description. Consider the challenge for members of the Supreme Court, whose job, as Justice Oliver Wendell Holmes Jr. observed, is to protect the First Amendment's guarantee of "freedom for the thought that we hate." That's pretty strong dissonance to overcome, although most judges imagine that they are up to the challenge. But according to a study of 4,519 votes by Supreme Court justices in over five hundred cases between 1953 and 2011, the justices were more likely to support freedom of speech for speakers whose speech they agreed with; conservative members of the Roberts court ruled in favor of conservative speakers about 65 percent of the time and liberal speakers about 21 percent. The gap for liberal justices was not as great, more like 10 percent, but they too were more likely to vote in support of speakers whose political philosophy they shared.[5]

We believe our own judgments are less biased and more independent than those of others partly because we rely on introspection to tell us what we are thinking and feeling, but we have no way of knowing what others are truly thinking.[6] And when we look into our souls and hearts, the need to avoid dissonance assures us that we have only the best and most honorable of motives. We take our own involvement in an issue as a source of accuracy and enlightenment ("I've felt strongly about gun control for years, therefore I know what I'm talking about"), but we regard such personal feelings on the part of others who hold different views as a source of bias ("She can't possibly be impartial about gun control because she's felt strongly about it for years").

All of us are as unaware of our blind spots as fish are unaware of the water they swim in, but those who swim in the waters of privilege have a particular motivation to remain oblivious. When Marynia Farnham achieved fame and fortune during the 1940s and 1950s by advising women to stay at home and raise children or risk frigidity, neuroses, and a loss of femininity, she saw no inconsistency (or irony) in the fact that she was privileged to be a physician who was not staying at home raising her own two children. When affluent people speak of the underprivileged, they rarely thank their lucky stars that they are privileged, let alone consider that they might be overprivileged. Privilege is their blind spot.[7] It is invisible and they don't think twice about it; they justify their social position as something they are entitled to. In one way or another, all of us are blind to whatever privileges life has handed us, even if those privileges are temporary. Most people who normally fly in an airline's main cabin regard the privileged people in business and first class as wasteful snobs, if enviable ones. Imagine paying all that extra money for a mere six-hour flight! But as soon as they are the ones paying for the business seats, that attitude vanishes, replaced by a self-justifying mixture of pity and disdain for their fellow passengers forlornly trooping past them into steerage.

Drivers cannot avoid having blind spots in their field of vision, but good drivers are aware of them; they know they had better be careful backing up and changing lanes if they don't want to crash into fire hydrants and other cars. Our innate biases are, as two legal scholars put it, "like optical illusions in two important respects — they lead us to wrong conclusions from data, and their apparent rightness persists even when we have been shown the trick."[8] We cannot avoid our psychological blind spots, but if we are unaware of them, we may become unwittingly reckless, crossing ethical lines and making foolish decisions. Introspection alone will not help our vision, because it will simply confirm our self-justifying beliefs that we, personally, cannot be co-opted or corrupted and that our dislikes or hatreds of other groups are not irrational but reasoned and legitimate. Blind spots enhance our pride and activate our prejudices.

The Road to St. Andrews

The greatest of faults, I should say, is to be conscious of none.

— *Thomas Carlyle, historian and essayist*

When *New York Times* editorial writer Dorothy Samuels learned that Tom DeLay, former leader of the House Republicans, had accepted a trip to the legendary St. Andrews golf course in Scotland from Jack Abramoff, a corrupt lobbyist then under investigation, she expressed her perplexity. "I've been writing about the foibles of powerful public officials for more years than I care to reveal without a subpoena," she wrote, "and I still don't get it: why would someone risk his or her reputation and career for a lobbyist-bestowed freebie like a vacation at a deluxe resort?"[9]

Why? Dissonance theory gives us the answer: one step at a time. Although there are plenty of unashamedly corrupt politicians who sell their votes to the largest campaign contributors, most politicians, thanks to their blind spots, believe they are incorruptible. When they first enter politics, they accept lunch with a lobbyist because, after all, that's how politics works and it's an efficient way to get information about a pending bill, isn't it? "Besides," the politician says, "lobbyists, like any other citizens, are exercising their right to free speech. I only have to listen; I'll decide how to vote on the basis of whether my party and constituents support this bill and on whether it is the right thing to do for the American people."

However, once you accept the first small inducement and justify it that way, you have started your slide down the pyramid. If you had lunch with a lobbyist to talk about that pending legislation, why not talk things over on the local golf course? What's the difference? It's a nicer place to have a conversation. And if you talked things over on the local course, why not accept a friendly offer to go to a better course to play golf with him or her — to, say, St. Andrews in Scotland? What's wrong with that? By the time the politician is at the bottom of the pyramid, having accepted and justified ever-larger inducements, the public is screaming, "What's *wrong* with that? Are you kidding?" At one level, the politician is not kidding. Dorothy Samuels is right: Who would jeopardize a career and reputation for a trip to Scotland? No one, if that was the first offer, but many of us would if that offer had been preceded by several smaller ones that we had accepted. Pride — when followed by self-justification — paves the road to Scotland.

Conflict of interest and politics are synonymous, and we all understand the cozy collaborations that politicians forge to preserve their own power at the expense of the common welfare. It's harder to see that exactly the same process affects judges, scientists, physicians, and other professionals who pride themselves on their ability to be intellectually independent for the sake of justice, scientific advancement, or public health. Their training and culture promote the core value of impartiality, so most people in these fields become indignant at the mere suggestion that financial or personal interests could contaminate their work. Their professional pride makes them see themselves as being above such matters. No doubt some are, just as, at the other extreme, some judges and scientists are flat-out dishonest, corrupted by ambition or money. In between the extremes of rare integrity and blatant dishonesty are the great majority who, being human, have all the blind spots the rest of us have. Unfortunately, they are also more likely to think they don't, which makes them even more vulnerable to being hooked.

Once upon a time, most scientists ignored the lure of commerce. When Jonas Salk was questioned in 1954 about whether he would be patenting his polio vaccine, he replied, "Could you patent the sun?" How charming and yet how naive his remark seems today; imagine handing over your discovery to the public interest without keeping a few million bucks for yourself. The culture of science valued the separation of research and commerce, and universities maintained a firewall between them. Because scientists got their money from the government or independent funding institutions, they were more or less free to spend years investigating a problem that might or might not pay off, either intellectually or practically. A scientist who went public and profited from his or her discoveries was regarded with suspicion, even disdain. "It was once considered unseemly for a biologist to be thinking about some kind of commercial enterprise while at the same time doing basic research," said bioethicist and scientist Sheldon Krimsky.[10] "The two didn't seem to mix. But as the leading figures of the field of biology began intensively finding commercial outlets and get-rich-quick schemes, they helped to change the ethos of the field. Now it is the multivested scientists who have the prestige."

The critical turning point occurred in 1980, when the Supreme Court ruled that patents could be issued on genetically modified bacteria independent of the process of development. That meant that you could get a patent for discovering a virus, altering a plant, isolating a gene, or modifying any other living organism as a "product of manufacture." The gold rush was on — the scientists' road to St. Andrews. Before long, many professors of molecular biology were serving on the advisory boards of biotechnology corporations and owned stock in companies selling products based on their research. Universities seeking new sources of revenue began establishing intellectual-property offices and providing incentives for faculty who patented their discoveries. Throughout the 1980s, the ideological climate shifted from one in which science was valued for its own sake or for the public interest to one in which science was valued for the profits it could generate in the private interest. Major changes in tax and patent laws were enacted, federal funding of research declined sharply, and tax benefits created a steep rise in funding from industry. The pharmaceutical industry was deregulated, and within a decade it had become one of the most profitable businesses in the United States.[11]

And then scandals involving conflicts of interest on the part of researchers and physicians began to erupt. Big Pharma was producing new, lifesaving drugs but also drugs that were unnecessary at best and risky at worst; more than three-fourths of all drugs approved between 1989 and 2000 offered only minor improvements over existing medications, cost nearly twice as much, and had higher risks.[12] By 1999, seven major drugs, in-

cluding Rezulin and Lotronex, had been taken off the market
for safety reasons. None had been necessary to save lives (one
was for heartburn, one a diet pill, one a painkiller, one an an-
tibiotic) and none was better than older, safer drugs. Yet these
seven drugs were responsible for 1,002 deaths and thousands of
troubling complications.[13] In 2017, researchers at the Yale School
of Medicine reported that nearly one-third of all new medications approved by the FDA between 2001 and 2010 had major
safety issues that were not apparent until they had been on the
market for an average of four years. Among the drugs withdrawn
were Bextra, an anti-inflammatory medication; Zelnorm, for ir-
ritable bowel syndrome; and Raptiva, for psoriasis. The first two
increased cardiovascular risk, and the third increased the risk
of a rare and fatal brain infection. Seventy-one of the 222 approved drugs were withdrawn, required a "black box" warning
about side effects, or warranted an announcement about newly
identified risks. These risks were greatest for antipsychotic medications, biologics, and drugs that had been granted "accelerated
approval."[14]

The public has reacted to such news not only with the anger
they are accustomed to feeling toward dishonest politicians but
also with dismay and surprise: How can scientists and physicians possibly promote a drug they know is harmful? Can't they
see that they are selling out? How can they justify what they are
doing? Certainly some investigators, like some politicians, are
corrupt and know exactly what they are doing. They are doing
what they were hired to do: getting results that their employers
want and suppressing results that their employers don't want to

hear about, as tobacco-company researchers did for decades. But at least public-interest groups, watchdog agencies, and independent scientists can eventually blow the whistle on bad or deceptive research. The greater danger to the public comes from the self-justifications of well-intentioned scientists and physicians who, because of their need to reduce dissonance, truly believe themselves to be above the influence of their corporate funders. Yet, like a plant turning toward the sun, they turn toward the interests of their sponsors without even being aware that they are doing so.

How do we know this? One way is through experimental studies that assess an expert's judgment and determine whether that judgment changes depending on who is paying for it. In one such experiment, researchers paid 108 forensic psychologists and psychiatrists the going rate to review four identical case files of actual sexual offenders and, using the same validated measures of risk assessment, offer their opinions on whether these men were more or less likely to reoffend. When experts use these measures in nonadversarial situations, their agreement is very high. But in this study, some of the experts were told they'd been hired by the defense; others were told they'd been hired by the prosecution, with the result that their assessments tilted toward their presumed employer: those who believed they were working for the prosecution assigned higher risk scores to offenders, and those who believed they were working for the defense assigned lower risk scores.[15]

Another way to measure the subtle effects of sponsorship is by comparing the results of studies funded independently and those funded by industry, which consistently reveal a funding bias.

- Two investigators selected 161 studies, all published during the same six-year span, of the possible risks to human health of four chemicals. Of the studies funded by industry, only 14 percent found harmful effects on health; of those funded independently, fully 60 percent found harmful effects.[16]

- A researcher examined more than 100 controlled clinical trials designed to determine the effectiveness of a new medication over older ones. Of those favoring the traditional drug, 13 percent had been funded by drug companies and 87 percent by nonprofit institutions.[17]

- Two Danish investigators examined 159 clinical trials that had been published between 1997 and 2001 in the *British Medical Journal,* where authors are required to declare potential conflicts of interest. The researchers could therefore compare studies in which the investigators had declared a conflict of interest with those in which there was none. The findings were "significantly more positive toward the experimental intervention" (i.e., the new drug compared to an older one) when the study had been funded by a for-profit organization.[18]

If most of the scientists funded by industry are not consciously cheating, what is causing the funding bias? Clinical trials of new drugs are complicated by many factors, including length of treatment, severity of the patients' disease, side effects, dosages, and variability in the patients being treated. The interpretation of results is rarely clear and unambiguous; that is why all scientific studies require replication and refinement and why most findings are open to legitimate differences of interpretation. If you are an impartial scientist and your research turns up an ambiguous but worrisome finding about your new drug, perhaps a slightly increased risk of heart attack or stroke, you might say, "This is troubling; let's investigate further. Is this increased risk a fluke, was it due to the drug, or were the patients unusually vulnerable?"

However, if you are motivated to show that your new drug is effective and better than older drugs, the better to keep your funding and your sponsor's approval, you will be inclined to downplay your misgivings and resolve the ambiguity in the company's favor. You will also be unconsciously motivated to seek only confirming evidence for your hypothesis — "It's nothing. There's no need to look further." "Those patients were already quite sick, anyway." "Let's assume the drug is safe until proven otherwise." This was the reasoning of the Merck-funded investigators who had been studying the company's multibillion-dollar painkiller Vioxx before evidence of the drug's risks was produced by independent scientists.[19]

In 1998, a team of scientists reported in the distinguished medical journal the *Lancet* that they had found a positive correlation between autism and the MMR (measles, mumps, rubella) vaccine. *Boom* — the announcement generated enormous fear and put scientists, physicians, and parents at the top of the pyramid with this decision: Should we stop vaccinating children? Thousands of parents stepped off in the direction of "yes," relieved that they now knew the reason for their children's autism or reassured that they had a way to prevent it.

Six years later, ten of the thirteen scientists involved in this study retracted that particular result and revealed that the lead author, Andrew Wakefield, had had a conflict of interest he had failed to disclose to the journal: he was conducting research on behalf of lawyers representing parents of autistic children. Wakefield had been paid more than eight hundred thousand dollars to determine whether there were grounds for pursuing legal action, and he gave the study's affirmative answer to the lawyers before publication. "We judge that all this information would have been material to our decision-making about the paper's suitability, credibility, and validity for publication," wrote Richard Horton, editor of the *Lancet*.[20]

Wakefield, however, did not sign the retraction and could not see a problem. "Conflict of interest," he wrote in his defense, "is created when involvement in one project potentially could, or actively does, interfere with the objective and dispassionate assessment of the processes or outcomes of another project. We cannot accept that the knowledge that affected children were later to pursue litigation, following their clinical referral and investigation, influenced the content or tone of [our earlier] paper . . . We emphasise that this was not a scientific paper but a clinical report."[21] Oh. It wasn't a scientific paper anyway.

No one knows Andrew Wakefield's real motives or thoughts about his research. But we suspect that he, like Stanley Berent in our opening story, convinced himself that he was acting honorably, that he was doing good work, and that he was uninfluenced by having been paid eight hundred thousand dollars by the lawyers. Unlike truly independent scientists, however, he had no incentive to look for disconfirming evidence of a correlation between vaccines and autism and many incentives to overlook other explanations. In fact, there is no causal relationship between autism and thimerosal, the preservative in the vaccines that was the supposed cause (thimerosal was removed from the vaccines in 2001, with no attendant decrease in autism rates). The apparent correlation was coincidental, a result of the fact that autism is typically diagnosed in children at the same age they are vaccinated.[22] As of 2019, more than a dozen large-scale, peer-reviewed studies, including a Danish project involving more than 650,000 children, had found no relationship between the MMR vaccine and autism.

And did the thousands of parents who had started their slide down the pyramid by deciding there *was* a relationship exclaim in relief, "Thank God for this helpful information"? Anyone who has been keeping up with the nationwide effort by some parents to block required vaccinations for their children knows the answer. Having spent six years justifying the belief that thimerosal was the agent responsible for their children's autism or other diseases, these parents rejected the research showing that it wasn't. They also rejected statements in favor of vaccination from the Centers for Disease Control and Prevention, the Food and Drug Administration, the National Institutes of Medicine, the World Health Organization, and the American Academy of Pediatrics. Faced with the dissonance between "I'm a good parent and know what's best for my child" and "Those organizations tell me I made a decision that could harm my child," what do they choose to believe? It's a no-brainer. "What do those scientists know, anyway," they say.

And that is how the "vaccinations cause autism" scare created tragic and lingering effects. A major epidemiological study found that vaccination programs for children have prevented more than a hundred million cases of serious contagious diseases since 1924 and saved between three and four million lives. But when some parents stopped vaccinating their children, rates of measles and whooping cough began to rise. The worst whooping cough epidemic since 1959 occurred in 2012, with 38,000 cases reported nationwide, and 2019 saw the greatest number of measles cases in twenty-five years — more than 1,250. This number represented a huge setback for public health, given that measles was declared eliminated in the United States in 2000. "Americans have witnessed an increase in hospitalizations and deaths from diseases like whooping cough, measles, mumps, and bacterial meningitis," writes Paul Offit, chief of the Division of Infectious Diseases and director of the Vaccine Education Center at the Children's Hospital of Philadelphia, "because some parents have become more frightened by vaccines than by the diseases they prevent."[23]

We noted in chapter 1 that people often hold on to a belief long after they know rationally that it's wrong, and this is especially true if they have taken many steps down the pyramid in support of that wrong belief. By then, getting information that contradicts a strong belief may actually backfire, causing the person to hold on to the incorrect belief even more firmly. Brendan Nyhan and his colleagues gave a nationally representative sample of parents various kinds of scientific information assuaging their worries about vaccines: information about disease risks, a dramatic story of what can happen if a child is not vaccinated, even tragic images of sick children. The parents who had had mixed or negative feelings toward vaccines actually became *less* likely to say they would vaccinate their children. They were persuaded that vaccines didn't cause autism, but they came up with other concerns or vague discomforts to justify their reluctance to vaccinate.[24] (Nyhan got the same results with people who didn't get flu shots because they wrongly believed the vaccine gave you the flu.)

That is the lingering legacy of self-justification, because most of the anti-vaccine alarmists have never said, "We were wrong, and look at the harm we caused." Andrew Wakefield, whose license was revoked by British medical authorities, stands by his view that vaccines cause autism. "I will not be deterred," he said in a press release. "This issue is far too important."[25] In 2015, following an extensive outbreak of measles that started at Disneyland, Barbara Loe Fisher, president of an anti-vaccine organization that spreads misinformation and combats efforts to ensure that children are vaccinated, said that all the concern was simply "hype," designed to cover up vaccine failures. Her group is located, we assume, in Fantasyland.[26]

The Gift That Keeps on Giving

Physicians, like scientists, want to believe their integrity cannot be compromised. Yet every time physicians accept a fee or other incentive for performing certain tests and procedures, for channeling some of their patients into clinical trials, or for prescribing a new, expensive drug that is not better or safer than an older one, they are balancing their patients' welfare against their own financial concerns. Their blind spot helps them tip the balance in their own favor, and then justify it: "If a pharmaceutical company wants to give us pens, notepads, calendars, lunches, honoraria, or small consulting fees, why not? We can't be bought by trinkets and pizzas." According to surveys, physicians regard small gifts as being ethically more acceptable than large gifts. The American Medical Association agrees, approving of gift-taking from pharmaceutical representatives as long as no single gift is worth much more than a hundred dollars. The evidence shows, however, that most physicians are influenced even more by small gifts than by big ones.[27]

Drug companies know this. A national random-sample survey of nearly three thousand primary-care physicians and specialists found that 84 percent reported having received some form of compensation from the pharmaceutical industry — drug samples, food and beverages, reimbursements, payments for services.[28] According to the Centers for Medicare and Medicaid Services, in a five-month period, from August to December of 2013, pharmaceutical companies and device makers paid a total of $3.5 billion to health-care professionals and teaching hospitals, an amount that included some $380 million in speaking and consulting fees to 546,000 individual physicians — and that early estimate proved to be about $1 billion short.[29] Some of those doctors were getting more than half a million dollars for their services, but the great majority were getting office trinkets, paid junkets, "continuing medical education" programs (where the only "education" is about the drug company's new medication), and "nonaccredited training."

The reason Big Pharma spends so much on small gifts as well as the big ones is well known to marketers, lobbyists, and social psychologists: being given a gift evokes an implicit desire to reciprocate. The Fuller Brush salespeople understood this principle decades ago when they pioneered the foot-in-the-door technique: Give a housewife a little brush as a gift, and she won't slam the door in your face. And once she hasn't slammed the door in your face, she will be more inclined to invite you in, and eventually to buy your expensive brushes. Robert Cialdini, who has spent many years studying influence and persuasion techniques, systematically observed Hare Krishna advocates raise money at airports.[30] Asking weary travelers for a donation wasn't working; the requests just made the travelers mad at them. And so the Krishnas came up with a better idea: They would approach a target traveler and press a flower into his hands or pin a flower to his jacket. If the target refused the flower and tried to give it back, the Krishna would demur and say, "It is our gift to you." Only then would the Krishna ask for a donation. This time the request was likely to be granted, because the gift of the flower had established a feeling of indebtedness and obligation in the traveler. How to repay the gift? With a small donation . . . and perhaps the purchase of a charming, overpriced edition of the Bhagavad Gita.

Were the travelers aware of the power of reciprocity to affect their behavior? Not at all. But once reciprocity kicks in, self-justification will follow: "I've always wanted a copy of the Bhagavad Gita; what is it, exactly?" The power of the flower is unconscious. "It's only a flower," the traveler says. "It's only a pizza," the medical resident says. "It's only a small donation for an educational symposium," the physician says. Yet the power of the flower is one reason that the amount of contact doctors have with pharmaceutical representatives is positively correlated with the cost of the drugs the doctors later prescribe. "That rep has been awfully persuasive about that new drug; I might as well try it; my patients might do well on it." Once you take the gift, no matter how small, the process starts. You will feel the urge to give something back, even if it's only, at first, your attention, your willingness to listen, your sympathy for the giver. Eventually, you will become more willing to give your prescription, your ruling, your vote. Your behavior changes, but, thanks to blind spots and self-justification, your view of your intellectual and professional integrity remains the same. A friend of ours was given a prescription for a drug that had a long list of cautions. When she sought out an independent website that noted that all the research on this drug was done by the pharmaceutical company that developed it, she pointed this out to her doctor. He said, "What difference does that make?"

Carl Elliott, a bioethicist and philosopher who also has an MD, has written extensively about the ways that small gifts entrap their recipients. His brother Hal, a psychiatrist, told him how he ended up on the speakers bureau of a large pharmaceutical company: First they asked him to give a talk about depression to a community group. Why not? he thought; it would be a public service. Next they asked him to speak on the same subject at a hospital. Next they began making suggestions about the content of his talk, urging him to speak not about depression but about antidepressants. Then they told him they could get him on a national speaking circuit, "where the real money is." Then they asked him to lecture about their own new antidepressant. Looking back, Hal told his brother:

> It's kind of like you're a woman at a party, and your boss says to you, "Look, do me a favor: be nice to this guy over there." And you see the guy is not bad-looking, and you're unattached, so you say, "Why not? I can be nice." Soon you find yourself on the way to a Bangkok brothel in the cargo hold of an unmarked plane. And you say, "Whoa, this is not what I agreed to." But then you have to ask yourself: "When did the prostitution actually start? Wasn't it at that party?"[31]

Nowadays, even professional ethicists are going to the party; the watchdogs are being tamed by the foxes they were trained to catch. Pharmaceutical and biotechnology industries are offering consulting fees, contracts, and honoraria to bioethicists, the very people who write about, among other things, the dangers of conflicts of interest between physicians and drug companies. Carl Elliott described his colleagues' justifications for taking the money. "Defenders of corporate consultation often bristle at the suggestion that accepting money from industry compromises their impartiality or makes them any less objective a moral critic," he wrote. "'Objectivity is a myth,' [bioethicist Evan] DeRenzo told me, marshaling arguments from feminist philosophy to bolster her cause. 'I don't think there is a person alive who is engaged in an activity who has absolutely no interest in how it will turn out.'" There's a clever dissonance-reducing claim for you — "Perfect objectivity is impossible anyway, so I might as well accept that consulting fee."

Thomas Donaldson, director of the ethics program at the Wharton School, justified this practice by comparing ethics consultants to independent accounting firms that a company might hire to audit their finances. Why not audit their ethics? This stab at self-justification didn't get past Carl Elliott either. "Ethical analysis does not look anything like a financial audit," he says. An accountant's transgression can be detected and verified, but how do you detect the transgressions of an ethics consultant? "How do you tell the difference between an ethics consultant who has changed her mind for legitimate reasons and one who has changed her mind for money? How do you distinguish between a consultant who has been hired for his integrity and one who has been hired because he supports what the company plans to do?"[32] Still, Elliott says wryly, perhaps we can be grateful that the AMA's Council on Ethical and Judicial Affairs designed an initiative to educate doctors about the ethical problems involved in accepting gifts from the drug industry. That initiative was funded by $590,000 in gifts from Eli Lilly and Company, GlaxoSmithKline, Pfizer, the U.S. Pharmaceutical Group, Astra-Zeneca Pharmaceuticals, the Bayer Corporation, Procter and Gamble, and Wyeth-Ayerst Pharmaceutical.

A Slip of the Brain

Al Campanis was a very nice man, even a sweet man, but also a flawed man who made one colossal mistake in his

81 years on earth — a mistake that would come to define him forevermore.

— *Mike Littwin, sportswriter*

On April 6, 1987, *Nightline* devoted its whole show to the fortieth anniversary of Jackie Robinson's Major League debut. Ted Koppel interviewed Al Campanis, general manager of the Los Angeles Dodgers, who had been part of the Dodger organization since 1943 and who had been Robinson's teammate on the Montreal Royals in 1946. That year, Campanis punched a bigoted player who had insulted Robinson and, subsequently, championed the admission of black players into Major League Baseball. And then, in talking with Koppel, Campanis put his brain on autopilot. Koppel asked Al, an old friend of Jackie Robinson's, why there were no black managers, general managers, or owners in baseball. Campanis was evasive at first — you have to pay your dues by working in the minors; there's not much pay while you're working your way up — but Koppel pressed him:

KOPPEL: Yeah, but you know in your heart of hearts . . . you know that that's a lot of baloney. I mean, there are a lot of black players, there are a lot of great black baseball men who would dearly love to be in managerial positions, and I guess what I'm really asking you is to, you know, peel it away a little bit. Just tell me why you think it is. Is there still that much prejudice in baseball today?

CAMPANIS: No, I don't believe it's prejudice. I truly believe that they may not have some of the necessities to be, let's say, a field manager, or perhaps a general manager.

KOPPEL: Do you really believe that?

CAMPANIS: Well, I don't say that all of them, but they certainly are short. How many quarterbacks do you have? How many pitchers do you have that are black?

Two days after this interview and the public uproar it caused, the Dodgers fired Campanis. A year later, he said he had been "wiped out" when the interview took place and therefore not entirely himself.

Who was the real Al Campanis? A bigot or a victim of political correctness? Neither. He was a man who liked and respected the black players he knew, who defended Jackie Robinson when doing so was neither fashionable nor expected, *and* who had a blind spot: He thought that black men were capable of being great players but weren't smart enough to be managers. And in his heart of hearts, he told Koppel, he didn't see what was wrong with that attitude; "I don't believe it's prejudice," he said. Campanis was not lying or being coy. But, as general manager, he was in a position to recommend the hiring of a black manager, and his blind spot kept him from even considering that possibility.

Just as we can identify hypocrisy in everyone but ourselves, just as it's obvious that others can be influenced by money but not ourselves, so we can see prejudices in everyone but ourselves. Thanks to our ego-preserving blind spots, we cannot possibly have a prejudice, which is an irrational or mean-spirited feeling about all members of another group. Because we are not irrational or mean-spirited, any negative feelings we have about another group are justified; our dislikes are rational and well founded. It's the other group's negative feelings we need to suppress. Like the

Hasids pounding on the Unprejudiced door at the Museum of Tolerance, we are blind to our own prejudices.

Prejudices emerge from the disposition of the human mind to perceive and process information in categories. *Categories* is a nicer, more neutral word than *stereotypes,* but it's the same thing. Cognitive psychologists view stereotypes as energy-saving devices that allow us to make efficient decisions on the basis of past experiences; they help us quickly process new information, retrieve memories, understand real differences between groups, and predict, often with considerable accuracy, how others will behave or think.[33] We wisely rely on stereotypes and the quick information they give us to avoid danger, approach possible new friends, choose one school or job over another, or decide that *that* person across this crowded room will be the love of our lives.

That's the upside. The downside is that stereotypes flatten out differences within the category we are looking at and exaggerate differences between categories. Red Staters and Blue Staters often see each other as nonoverlapping categories, but plenty of Kansans do want evolution taught in their schools, and plenty of Californians oppose any kind of gun control. All of us recognize variation within our own gender, party, ethnicity, or nation, but we are inclined to generalize about people in other categories and lump them all together as *them*. This habit starts awfully early. Social psychologist Marilynn Brewer, who studied the nature of stereotypes for many years, reported that her daughter once returned from kindergarten complaining that "boys are crybabies."[34] The child's evidence was that she had seen two boys crying on their first day away from home. Brewer, ever the scientist, asked whether there hadn't also been little girls who cried. "Oh yes," said her daughter. "But only *some* girls cry. I didn't cry."

Brewer's little girl was already dividing the world into us and them. *Us* is the most fundamental social category in the brain's organizing system, and the concept is hardwired. Even the plural pronouns *us* and *them* are powerful emotional signals. In one experiment in which participants believed their verbal skills were being tested, a nonsense syllable such as *xeh, yof, laj,* or *wuh* was randomly paired with an in-group word (*us, we,* or *ours*), an out-group word (*them, they,* or *theirs*), or, for a control measure, another pronoun (such as *he, hers,* or *yours*). All participants then had to rate the syllables on how pleasant or unpleasant they were. You might wonder why anyone would have an emotional feeling toward a nonsense word like *yof* or think *wuh* was cuter than *laj.* Yet participants liked the nonsense syllables more when they were linked with in-group words than with any other word.[35] Not one of them guessed why; not one was aware of how the words had been paired.

As soon as people have created a category called *us,* however, they invariably perceive everybody who isn't in it as *not-us.* The specific content of *us* can change in a flash: It's us sensible Midwesterners against you flashy coastal types; it's us Prius owners against you gas-guzzling-SUV owners; it's us Boston Red Sox fans against you Los Angeles Angels fans (to pick a random example that happens to describe your two authors during baseball season). "Us-ness" can be manufactured in a minute in the laboratory, as Henri Tajfel and his colleagues demonstrated in a classic experiment with British schoolboys.[36] Tajfel showed the boys slides with varying numbers of dots on them and asked them to guess how many dots there were. He arbitrarily told some of them that they were overestimators and others that they were underestimators and then asked all the boys to work on another task. In this phase, they had a chance to give points to other boys identified as overestimators or underestimators. Although each boy worked alone in his cubicle, almost every single one assigned more points to boys he thought were like him, an overestimator or an underestimator. As the boys emerged from their rooms, the other kids asked them, "Which were you?" The answers received cheers from those like them and boos from the others.

Obviously, certain categories of *us* are more crucial to our identities than the kind of cars we drive or the number of dots we estimate on a slide — gender, sexuality, religion, politics, ethnicity, and nationality, for starters. Without feeling attached to groups that give our lives meaning, identity, and purpose, we would suffer the intolerable sensation that we were loose marbles rattling around in a random universe. Therefore, we will do what it takes to preserve these attachments. Evolutionary psychologists argue that ethnocentrism — the belief that your own culture, nation, or religion is superior to all others — aids survival by strengthening your bonds to your primary social groups and thus increasing your willingness to work, fight, and occasionally die for them. When things are going well, most of us feel pretty tolerant of other cultures and religions — and even of the other sex! — but when we are angry, anxious, or threatened, our blind spots are automatically activated. *We* have the human qualities of intelligence and deep emotions, but *they* are dumb, *they* are crybabies, *they* don't know the meaning of love, shame, grief, or remorse.[37]

The very act of thinking that *they* are not as smart or reasonable as *we* are makes us feel closer to others who are like us. But, just as crucially, it allows us to justify how we treat *them*. Most people assume that stereotyping causes discrimination; Al Campanis, believing that blacks lack the "necessities" to be managers, refused to hire one. But the theory of cognitive dissonance shows that the path between attitudes and action runs in both directions. Often it is discrimination that evokes the self-justifying stereotype; Al Campanis, lacking the will or guts to convince the Dodger organization to hire a black manager, justified his failure to act by telling himself that blacks couldn't do the job anyway. In the same way, if we have enslaved members of another group, deprived them of decent educations or jobs, kept them from encroaching on our professional turfs, or denied them their human rights, then we invoke stereotypes about them to justify our actions. By persuading ourselves that they are unworthy, unteachable, incompetent, inherently math-challenged, immoral, sinful, stupid, or even subhuman, we avoid feeling guilty or unethical about how we treat them. And we certainly avoid feeling that we are prejudiced. Why, we even like some of those people, as long as they know their place, which, it goes without saying, is not here in our club, our university, our job, our neighborhood. In short, we use stereotypes to justify behavior that would otherwise make us feel bad about the kind of people we are or the kind of country we live in.

But given that thinking in categories is a universal feature of the mind, why do only some people hold bitter, passionate prejudices toward other groups? Al Campanis was not prejudiced in terms of his having a strong emotional antipathy toward blacks;

we suspect he could have been argued out of his notion that black players could not be good managers. A stereotype might bend or even shatter under the weight of disconfirming information, but the hallmark of *prejudice* is that it is impervious to reason, experience, and counterexample. In his timeless book *The Nature of Prejudice,* written in 1954, social psychologist Gordon Allport described the responses characteristic of a prejudiced man when confronted with evidence contradicting his beliefs:

MR. X: The trouble with Jews is that they only take care of their own group.

MR. Y: But the record of the Community Chest campaign shows that they give more generously, in proportion to their numbers, to the general charities of the community, than do non-Jews.

MR. X: That shows they are always trying to buy favor and intrude into Christian affairs. They think of nothing but money; that is why there are so many Jewish bankers.

MR. Y: But a recent study shows that the percentage of Jews in the banking business is negligible, far smaller than the percentage of non-Jews.

MR. X: That's just it; they don't go in for respectable business; they are only in the movie business or run night clubs.[38]

Allport nailed Mr. X's thought processes perfectly. Mr. X doesn't even try to respond to Mr. Y's evidence; he just slides along to another reason for his dislike of Jews. Once people have a prejudice, just as once they have a political ideology, they do not easily drop it, even if the evidence indisputably contradicts a core justification for it. Rather, they come up with another justification to preserve their belief or rationalize a course of action. Suppose our reasonable Mr. Y told you that insects were a great source of protein and that the sensational new chef at the Slugs and Bugs Diner is offering delicious entrées involving puréed caterpillars. Will you rush out to try this culinary adventure? If you have a prejudice against eating insects, probably not, even if this chef has made the front page of the *New York Times* Food section. You will, like the bigoted Mr. X, find another reason to justify it. "Ugh," you would tell Mr. Y, "insects are ugly and squishy." "Sure," he says. "Tell me again why you eat lobster and raw oysters?"

An acquired prejudice is hard to dislodge. As the great jurist Oliver Wendell Holmes Jr. said, "Trying to educate a bigot is like shining light into the pupil of an eye — it constricts." Most people will expend a lot of mental energy to avoid having to change their prejudices, often by waving away disconfirming evidence as "exceptions that prove the rule." (What would disprove the rule, we wonder.) The line "But some of my best friends are [X]," well deserving of the taunts it now gets, has persisted because it is such an efficient way of resolving the dissonance created when a prejudice runs headlong into an exception. When Elliot moved to Minneapolis years ago to teach at the University of Minnesota, a neighbor said to him, "You're Jewish? But you're so much nicer than . . ." She stopped. "Than what?" he asked. "Than what I expected," she finished lamely. By admitting that Elliot didn't fit her stereotype, she was able to feel open-minded and generous while maintaining her basic prejudice toward the whole category of Jews. In her mind, she was even paying him a compliment: "He's so much nicer than all those others of his . . . race."

Jeffrey Sherman and his colleagues have done a series of experiments that demonstrate the effort that highly prejudiced people are prepared to put into maintaining consonance between their prejudices and information that is inconsistent with it. They actually pay more attention to this inconsistent information than to consistent information, because, like Mr. X and the Minnesota neighbor, they need to figure out how to explain away the dissonant evidence. In one experiment, straight students were asked to evaluate a gay man, Robert, who was described as doing eight things that were consistent with the gay stereotype (e.g., he had studied interpretive dance) and eight things that were inconsistent (e.g., he had watched a football game one Sunday). Antigay participants twisted the evidence about Robert and later described him as being far more "feminine" than unbiased students did, thereby maintaining their prejudice. To resolve the dissonance caused by the inconsistent facts, they explained them away as being an artifact of the situation. Sure, Robert watched a football game, but only because his cousin Fred was visiting.[39]

These dissonance-reducing contortions occur in the world outside the lab all the time. Consider the lengths that some white supremacists go to upon learning that a potential ally is not 100 percent "white." Aaron Panofsky and Joan Donovan examined hundreds of posts on the website of the white nationalist group Stormfront to see how the organization counsels applicants who report the "upsetting news" that their DNA revealed some non-white or non-European ancestry. Stormfront's founders have absolutist rules of membership. They will admit only "non-Jewish people of wholly European descent. No exceptions," and they state that whiteness is determined genetically. But given that they want as many members as they can get, what are they to do with a would-be member whose DNA indicates nonwhite ancestry? They can reduce dissonance in two ways — the strict way and the flexible way. The strict way is to kick them out:

POST: Hello, got my DNA results and I learned today I am 61% European. I am very proud of my white race and my european roots. I know many of you are 'whitter' [*sic*] than me, I don't care, our goal is the same. I would like to do anything possible to protect our white race, our european roots and our white families.

RESPONSE: I've prepared you a drink. It's 61% pure water. The rest is potassium cyanide. I assume you have no objections to drinking it. . . . Cyanide isn't water, and YOU are not White.

But the researchers found that most Stormfront members, to increase their numbers, reduce dissonance and console worried would-be supporters by offering unscientific reasons why the results can't be trusted — "There are many ways of measuring whiteness, so stay with us"; "The tests' statistics were not interpreted accurately"; and the ever-popular Jewish-conspiracy theory: "Jews own those genetic-testing companies and we all know about their malevolent multicultural agenda." (There is "the fact that 23 and Me is Jewish controlled and it would not be surprising if all the others are too," wrote one member. "I think 23 and Me might be a covert operation to get DNA the Jews could then use to create bio-weapons for use against us.") Even people who reported evidence of black and Jewish ancestry, the two most despised ethnicities on Stormfront, got reassuring responses designed to explain away or minimize the dissonant results. One woman appealed "in a panic" to the community to help her interpret the evidence that her mother's DNA showed "PersianTurkishCaucasus 11%"; did that mean she was racially contaminated? Don't worry, said one respondent. Though the Caucasus population is Muslim today, it was "white originally" and the "Persians are Aryans."[40]

Stormfront members and other unapologetic white nationalists flaunt their prejudices. But most Americans who are prejudiced against a particular group know better than to announce that fact, given that many people live and work in environments where they can be slapped on the wrist, publicly humiliated, or sacked for saying anything that smacks of an ism. However, just as it takes mental effort to maintain a prejudice despite conflicting information, it also takes mental effort to suppress those negative feelings. Social psychologists Chris Crandall and Amy Eshelman, reviewing the huge research literature on prejudice, found that whenever people are emotionally depleted — when they are sleepy, frustrated, angry, anxious, drunk, or stressed — they become more willing to express their real prejudices toward another group. When Mel Gibson was arrested for drunk driving and launched into an anti-Semitic tirade, he claimed, in his inevitable statement of apology the next day, that "I said things that I do not believe to be true and which are despicable. I am deeply ashamed of everything I said . . . I apologize for any behavior unbecoming of me in my inebriated state." Translation:

"It wasn't me, it was the booze." Nice try, but the evidence shows clearly that while inebriation makes it easier for people to reveal their prejudices, it doesn't put those attitudes in their minds. Therefore, when people apologize by saying, "I don't really believe those things I said; I was [tired/worried/angry/drunk]" — or, as Al Campanis put it, "wiped out" — we can be pretty sure they really do believe it.

But most people are unhappy about believing it, and that creates dissonance: "I dislike those people" collides with an equally strong conviction that it is morally or socially wrong to say so. People who feel this dissonance, Crandall and Eshelman suggest, will eagerly reach for any self-justification that allows them to express their true beliefs yet continue to feel that they are moral and good. No wonder it is such a popular dissonance reducer. Even Donald Trump, with his rants against a long list of groups he dislikes (notably Latinos, Muslims, and disabled people), his promulgation of the "birther" lie that Barack Obama was not born in the United States, and his history of discriminatory treatment of African Americans, felt the need to assure the public via Twitter that "I am the least racist person you have ever met" and that "I don't have a Racist bone in my body!" "Justification," Crandall and Eshelman explain, "undoes suppression, it provides cover, and it protects a sense of egalitarianism and a nonprejudiced self-image."[41]

In one typical experiment, white students were told they would be inflicting an electric shock on another student, the learner, ostensibly as part of a study of biofeedback. The students working with a black learner initially gave lower-intensity shocks than students working with a white one, reflecting a desire, perhaps, to show they were not prejudiced. Then the students overheard the learner making derogatory comments about them, which, naturally, made them angry. Now, given another opportunity to inflict electric shock, the students who were working with a black learner administered higher levels of shock than students who were working with a white learner. The same result appears in studies of how English-speaking Canadians behave toward French-speaking Canadians, straights toward homosexuals, non-Jewish students toward Jews, and men toward women.[42] Participants successfully control their negative feelings under normal conditions, but as soon as they become angry or frustrated or when their self-esteem wobbles, they express their prejudice directly because now they can justify it: "I'm not a bad or prejudiced person, but, hey — he insulted me!"

In this way, prejudice is the energy of ethnocentrism. It lurks there, napping, until ethnocentrism summons it to do its dirty work, justifying the occasional bad things we good people want to do. In the nineteenth-century American West, Chinese immigrants were hired to work in the gold mines, potentially taking jobs from white laborers. The white-run newspapers fomented prejudice against them, describing the Chinese as "depraved and vicious," "gross gluttons," "bloodthirsty and inhuman." Yet only a decade later, when the Chinese were willing to accept the dangerous, arduous work of building the transcontinental railroad — work that white laborers were unwilling to undertake — public prejudice toward them subsided, replaced by the opinion that the Chinese were sober, industrious, and law-abiding. "They are equal to the best white men," said the railroad tycoon

Charles Crocker. "They are very trusty, very intelligent and they live up to their contracts." After the completion of the railroad, jobs again became scarce, and the end of the Civil War brought an influx of war veterans into an already tight job market. Anti-Chinese prejudice returned, with the press now describing the Chinese as "criminal," "conniving," "crafty," and "stupid."[43]

Prejudice justifies the ill treatment we inflict on others, and we want to inflict ill treatment on others because we don't like them. And why don't we like them? Because they are competing with us for jobs in a tough job market. Because their presence makes us doubt that ours is the one true religion. Because we want to preserve our positions of status, power, and privilege. Because our country is waging war against them. Because we are uncomfortable with their customs, especially their sexual customs, those promiscuous perverts. Because they refuse to assimilate into our culture. Because they are trying too hard to assimilate into our culture. Because we need to feel we are better than *somebody*.

By understanding prejudice as our self-justifying servant, we can better see why some prejudices are so hard to eradicate: They allow people to justify and defend their most important social identities — their "white" race, their religion, their gender, their sexuality — while reducing the dissonance between "I am a good person" and "I really don't like those people." Fortunately, we can also better understand the conditions under which prejudices diminish: when the economic competition subsides, when the truce is signed, when the profession is integrated, when *they* become more familiar and comfortable, when we stop seeing *them* as an undifferentiated mass and realize that they are as diverse a collection of individuals as we are.

• • •

"In normal circumstances," wrote Hitler's henchman Albert Speer in his memoirs, "people who turn their backs on reality are soon set straight by the mockery and criticism of those around them, which makes them aware they have lost credibility. In the Third Reich there were no such correctives, especially for those who belonged to the upper stratum. On the contrary, every self-deception was multiplied as in a hall of distorting mirrors, becoming a repeatedly confirmed picture of a fantastical dream world which no longer bore any relationship to the grim outside world. In those mirrors I could see nothing but my own face reproduced many times over."[44]

Our greatest hope of self-correction lies in making sure we are not operating in a hall of mirrors in which all we see are distorted reflections of our own desires and convictions. We need a few trusted naysayers in our lives, critics who are willing to puncture our protective bubble of self-justifications and yank us back to reality if we veer too far off. This is especially important for people in positions of power.

According to historian Doris Kearns Goodwin, Abraham Lincoln was one of the rare presidents who understood the importance of surrounding himself with people willing to disagree with him. Lincoln created a cabinet that included four of his political opponents, three of whom had run against him for the Republican nomination in 1860 and who felt humiliated, shaken, and angry to have lost to a relatively unknown backwoods lawyer: William H. Seward (whom Lincoln made secretary of state), Salmon P. Chase (secretary of the Treasury), and Edward Bates (attorney general). Although all shared Lincoln's goal of preserving the Union and ending slavery, this "team of rivals" (as Goodwin calls them) disagreed with one another furiously on how to do it.

Early in the Civil War, Lincoln was in deep trouble politically. He had to placate not only the Northern abolitionists who wanted escaped slaves emancipated but also the slave owners from border states like Missouri and Kentucky. These border states could have joined the Confederacy at any time, which would have been a disaster for the Union. As a result of the ensuing debates with his advisers, all of whom had different ideas about how to keep both sides in line, Lincoln could not delude himself that he had group consensus on every decision. He was able to consider alternatives and eventually enlist the respect and support of his erstwhile competitors.[45]

As long as we are convinced that we are completely objective, above corruption, and immune to prejudice, most of us from time to time will find ourselves on our own personal road to St. Andrews — and some of us will be on that plane to Bangkok. Jeb Stuart Magruder, whose entrapment in the political corruption of the Watergate scandal we described in the previous chapter, was blinded by his belief in the importance of doing whatever it took, even if that involved illegal actions, to defeat "them," Nixon's political enemies. But when he was caught, Magruder had the guts to face himself. It's a shocking, excruciating moment for anyone, like catching sight of yourself in a mirror and realizing that you've got a huge purple growth on your forehead. Magruder could have done what most of us would be inclined to do: Get some heavy makeup and say, "What purple growth?" But he resisted the impulse. In the final analysis, Magruder said, no one forced him or the others to break the law. "We could have objected to what was happening or resigned in protest," he wrote.[46] "Instead, we convinced ourselves that wrong was right, and plunged ahead.

"There is no way to justify burglary, wiretapping, perjury, and all the other elements of the cover-up . . . I and others rationalized illegal actions on the grounds of 'politics as usual' or 'intelligence gathering' or 'national security.' We were completely wrong, and only when we have admitted that and paid the public price of our mistakes can we expect the public at large to have much faith in our government or our political system."

3

Memory, the Self-Justifying Historian

What we . . . refer to confidently as memory . . . is really a form of storytelling that goes on continually in the mind and often changes with the telling.
— *William Maxwell, memoirist and editor*

Many years ago, during the Carter administration, the flamboyant novelist and media personality Gore Vidal was interviewed on the *Today* show by Tom Brokaw, the eminent TV journalist and host. According to Vidal, Brokaw said, "You've written a lot about bisexuality —" and Vidal cut him off, saying, "Tom, let me tell you about these morning shows. It's too early to talk about sex. Nobody wants to hear about it at this hour, or if they do, they are doing it. Don't bring it up." "Yeah, uh, but Gore, uh, you have written a lot about bisex —" Vidal interrupted again, saying that his new book had nothing to do with bisexuality and he'd rather talk about politics. Brokaw tried once more, and Vidal again declined to discuss the topic, saying, "Now let's talk about Carter . . . What is he doing with these Brazilian dictators pretending they are freedom-loving, democratic leaders?" And so the conversation turned to Carter for the rest of the interview. Several years later, when Brokaw became anchor of the *Nightly News, Time* did a feature on him and asked him about any especially difficult interviews he had conducted. Brokaw singled out the conversation with Gore Vidal: "I wanted to talk politics," Brokaw recalled, "and he wanted to talk about bisexuality."

It was a "total reversal," Vidal said, "to make me the villain of the story."[1]

Was it Tom Brokaw's intention to turn Gore Vidal into the villain of the story? Was Brokaw lying, as Vidal implied? That is unlikely. After all, Brokaw chose the story to tell the *Time* reporter; he could have selected any difficult interview in his long career rather than one that required him to embellish or lie. Indeed, for all he knew, the reporter would check the original transcript. Brokaw made the reversal of who-said-what unconsciously, not to make Vidal look bad but to make himself look good. It would have been unseemly for the new anchor of the *Nightly News* to have asked questions about bisexuality; better to believe (and remember) that he had always chosen the intellectual high road of politics.

When two people produce entirely different memories of the same event, observers usually assume that one of them is lying. Of course, some people do invent or embellish stories to manipulate or deceive their audiences (or sell books). But most of us, most of the time, are neither telling the whole truth nor intentionally deceiving. We aren't lying; we are self-justifying. All of us, as we tell our stories, add details and omit inconvenient facts; we give the tale a small, self-enhancing spin. That spin goes over so well that the next time we add a slightly more dramatic embellishment; we justify that little white lie as making the story better and clearer. Eventually the way we remember the event may bring us a far distance from what actually happened.

In this way, memory becomes our personal, live-in, self-justifying historian. Social psychologist Anthony Greenwald has described the self as being ruled by a "totalitarian ego" that ruthlessly destroys information it doesn't want to hear and, like all fascist leaders, rewrites history from the standpoint of the victor.[2] But whereas a totalitarian ruler rewrites history to put one over on future generations, the totalitarian ego rewrites history to put one over on itself. History is written by the victors, and when we write our own histories, we have the same goals as the conquerors of nations have: to justify our actions and make us look and feel good about ourselves and what we did or failed to do. If mistakes were made, memory helps us remember that they were made by someone else. If we were there, we were just innocent bystanders.

At the simplest level, memory smooths out the wrinkles of dissonance by enabling the confirmation bias to hum along, selectively causing us to forget discrepant, disconfirming information about beliefs we hold dear. If we were perfectly rational beings, we would try to remember smart, sensible ideas and not bother taxing our minds by remembering foolish ones. But dissonance theory predicts that we will conveniently forget good arguments made by an opponent, just as we forget foolish arguments made by our own side. A silly argument in favor of our own position arouses dissonance because it raises doubts about the wisdom of that position or the intelligence of the people who agree with it. Likewise, a sensible argument by an opponent arouses dissonance because it raises the possibility that the other side, God forbid, may be right or have a point we should take seriously. Because a silly argument on our side and a good argument on the other guy's side both arouse dissonance, the theory predicts that we will either not learn these arguments well or forget them quickly. And that is just what Edward Jones and Rika Kohler showed in a classic 1958 experiment on attitudes toward desegregation in North Carolina.[3] Each side tended to remember the plausible arguments agreeing with their own position and the implausible arguments agreeing with the opposing position; each side forgot the implausible arguments for their view and the plausible arguments for the opposition.

Naturally, some memories can be remarkably detailed and accurate. We remember first kisses and favorite teachers. We remember family stories, movies, dates, baseball stats, childhood humiliations and triumphs. We remember the central events of our life stories. But when we do misremember, our mistakes aren't random. The daily, dissonance-reducing distortions of memory help us make sense of the world and our place in it, protecting our decisions and beliefs. The distortion is even more powerful when it is motivated by the need to keep our self-concept consistent, by the wish to be right, by the need to preserve self-esteem, by the need to excuse failures or bad decisions, or by the need to find an explanation, preferably one safely in the past, of current problems.[4] Confabulation, distortion, and plain forgetting are the foot soldiers of memory, and they are summoned to the front lines when the totalitarian ego wants to protect us from the pain and embarrassment of actions we took that are dissonant with

our core self-images: "I did *that?*" That is why memory research-
ers love to quote Nietzsche: "'I have done that,' says my memory.
'I cannot have done that,' says my pride, and remains inexorable.
Eventually — memory yields."

The Biases of Memory

One of us (Carol) had a favorite children's book, James Thur-
ber's *The Wonderful O,* that she remembers her father giving her
when she was a child. "A band of pirates takes over an island and
forbids the locals to speak any word or use any object containing the letter *O*," Carol recalls. "I have a vivid memory of my father reading *The Wonderful O* and our laughing together at the
thought of shy Ophelia Oliver saying her name without its *O*s.
I remember trying valiantly, along with the invaded islanders, to
guess the fourth *O* word that must never be lost (after *love, hope,*
and *valor*), and my father's teasing guesses: Oregon? Orangutan?
Ophthalmologist? And then, not long ago, I found my first edition of *The Wonderful O.* It had been published in 1957, one
year after my father's death. I stared at that date in disbelief and
shock. Obviously, someone else gave me that book, someone else
read it to me, someone else laughed with me about 'Phelia 'Liver,
someone else wanted me to understand that the fourth *O* was
freedom. Someone lost to my recollection."

This small story illustrates three important things about
memory: how disorienting it is to realize that a vivid memory,
one full of emotion and detail, is indisputably wrong; how even
being absolutely, positively sure a memory is accurate does not
mean that it is; and how errors in memory support our cur-

rent feelings and beliefs. "I have a set of beliefs about my father," Carol observes, "the warm man he was, the funny and devoted dad who loved to read to me and take me rummaging through libraries, the lover of wordplay. So it was logical for me to assume — no, to *remember* — that he was the one who read me *The Wonderful O.*"

The metaphors of memory fit our times and technology. Centuries ago, philosophers compared memory to a soft wax tablet that would preserve anything imprinted on it. With the advent of the printing press, people began to think of memory as a library or perhaps a set of filing cabinets; events and facts could be stored for later retrieval if you could only find them in the damned card catalog. With the inventions of movies and tape recorders, people started thinking of memory as a video camera, clicking on at the moment of birth. Nowadays we think of memory in computer terms, and although some of us wish for more RAM, we assume that just about everything that happens to us is "saved." Your brain might not choose to display all those memories, but they are in there, just waiting for you to access them, bring them up on the screen, get out the popcorn, and watch.

These metaphors of memory are popular, reassuring, and wrong. Memories are not buried somewhere in the brain like bones at an archaeological site; you can't dig them up, perfectly preserved. We do not remember everything that happens to us; we select only highlights. If we didn't forget, our minds could not work efficiently, because they would be cluttered with mental junk — the temperature last Wednesday, a boring conversation on the bus, the price of peaches at the market yesterday. A very few people have a condition that allows them to remember just about everything, from a random fact like the weather on March 12, 1997, to public events to personal experiences, but this talent is not always the blessing it might appear. One woman with this ability described her memory as "non-stop, uncontrollable and totally exhausting" and "a burden."[5] Judicious pruning of memories is thus adaptive, and even people with extraordinary memories are not "recording" everything that happens to them as if on video.

Moreover, recovering a memory is not at all like retrieving a file or playing back a recording; it is like watching a few unconnected frames of a film and then figuring out what the rest of the scene must have been like. We may reproduce poetry, jokes, and other kinds of information by rote, but when we remember complex information, we shape it to fit it into a story line.

Because memory is reconstructive, it is subject to confabulation — confusing an event that happened to someone else with one that happened to you or coming to believe that you remember something that never happened. In reconstructing a memory, people draw on many sources. When you remember your fifth birthday party, you may have a direct recollection of your younger brother putting his finger in the cake and spoiling it for you, but you will also incorporate information that you got later from family stories, photographs, home videos, and birthday parties you've seen on television. You weave all these elements together into one integrated account. If someone hypnotizes you and regresses you to your fifth birthday party, you'll tell a lively story about it that will feel terribly real to you, but it will include many of those party details that never actually happened. After a while, you won't be able to distinguish your actual memory from subsequent information that crept in from elsewhere. That phenomenon is called "source confusion," otherwise known as the "where did I hear that?" problem.[6] Did I read it, see it, or did someone tell me about it?

Mary McCarthy made brilliant use of her understanding of confabulation in *Memories of a Catholic Girlhood,* which is a rare exception to the way most of us tell our stories. At the end of each chapter, McCarthy subjected her memories to the evidence for or against them, even when the evidence killed a good story. In "A Tin Butterfly," McCarthy vividly recalls the time her punitive uncle Myers and aunt Margaret, the relatives who took her and her brothers in when their parents died, accused her of stealing her younger brother's Cracker Jack prize, a tin butterfly. She hadn't, and a thorough household search failed to uncover it. But one night after dinner the butterfly was discovered under the tablecloth on the dining table, near Mary's place. Mary's uncle and aunt whipped her furiously for this alleged theft, he with a strop, she with a hairbrush, but the question of what had happened to the toy remained a mystery. Years later, when the siblings were grown and reminiscing together, they got to talking about the dreaded Uncle Myers. "It was then my brother Preston told me," McCarthy writes, "that on the famous night of the butterfly, he had seen Uncle Myers steal into the dining room from the den and lift the tablecloth, with the tin butterfly in his hand."

End of chapter. Fabulous! A dramatic ending, brilliantly told. And then McCarthy adds a postscript. As she was writing the story, she says, "I suddenly remembered that in college I had started writing a play on the subject. Could the idea that Uncle Myers put the butterfly at my place have been suggested to me by my teacher? I can almost hear her voice saying to me, excitedly: 'Your uncle must have done it!'" McCarthy called her brothers, but none of them recalled her version of events, including Preston, who did not remember either seeing Uncle Myers with the butterfly (he was only seven at the time) or claiming that he had said so the night of the family visit. "The most likely thing, I fear," McCarthy concludes, "is that I fused two memories" — the tale of the missing butterfly and the teacher's subsequent explanation of what might have happened.[7] And it made psychological sense: Uncle Myers's planting of the butterfly under the tablecloth was consonant with McCarthy's feelings about his overall malevolence and further justified her righteous indignation about being unfairly punished.

When most people write their memoirs or describe their past experiences, however, they don't do it the way Mary McCarthy did. They do it the way they would tell their stories to a therapist: "Doctor, here's what happened." They count on the listener not to say, "Oh, yeah? Are you sure it happened that way? Are you positive your mother hated you? Are you certain your father was such a brute? And while we're at it, let's examine those memories you have of your horrible ex. Any chance you have forgotten anything *you* did that might have been a tad annoying — say, that little affair you justified having with the lawyer from Bugtussle, Oklahoma?" On the contrary, we tell our stories in the confidence that the listener will not dispute them or ask for disconfirming evidence, which means we rarely have an incentive to scrutinize them for accuracy. You have memories about your father that are salient to you and that represent the man he was and the relationship you had with him. What have you forgotten? You remember that time when you were disobedient and he swatted you, and you are still angry that he didn't explain why he was disciplining you. But could you have been the kind of kid a father couldn't explain things to because you were impatient and impulsive and didn't listen? When we tell a story, we tend to leave ourselves out: My father did thus-and-such because of who he was, not because of the kind of kid I was. That's the self-justification of memory. And it is why, when we learn that a memory is wrong, we feel stunned, disoriented, as if the ground under us has shifted. In a sense, it has. It has made us rethink our own role in the story.

Every parent has been an unwilling player in the you-can't-win game. Require your daughter to take piano lessons, and later she will complain that you wrecked her love of the piano. Let your daughter give up lessons because she doesn't want to practice, and later she will complain that you should have forced her to keep going — why, now she can't play the piano at all. Require your son to go to Hebrew school in the afternoon, and he will blame you for having kept him from becoming another Hank Greenberg. Allow your son to skip Hebrew school, and he will later blame you for his not feeling more connected to his heritage. Betsy Petersen produced a full-bodied whine in her memoir *Dancing with Daddy,* blaming her parents for giving her swimming lessons, trampoline lessons, horseback-riding lessons, and tennis lessons but not ballet lessons. "The only thing I wanted, they would not give me," she wrote. Parent blaming is a popular and convenient form of self-justification because it allows people to live less uncomfortably with their regrets and imperfections. Mistakes were made, but only by my parents. Never mind that I raised hell about those lessons or stubbornly refused to take advantage of them. Memory thus minimizes our own responsibility and exaggerates theirs.

By far the most important distortions and confabulations of memory are those that serve to justify and explain our own lives. The mind, sense-making organ that it is, does not interpret our experiences as if they were separate shards of glass; it assembles them into a mosaic. From the distance of years, we see the mosaic's pattern. It seems tangible, unchangeable; we can't imagine how we could reconfigure those pieces into another design. But it is a result of years of telling our story, shaping it into a life narrative that is complete with heroes and villains, an account of how we came to be the way we are. Because that narrative is the way we understand the world and our place in it, it is bigger than the sum of its parts. If one part, one memory, is shown to be wrong, people have to reduce the resulting dissonance and even rethink the basic mental category: You mean Dad [Mom] wasn't such a bad [good] person after all? You mean Dad [Mom] was a complex human being? The life narrative may be fundamentally true; your father or mother might really have been hateful, or saintly. The problem is that when the narrative becomes a major source of self-justification, one the storyteller relies on to excuse mistakes and failings, memory becomes warped in its service. The storyteller remembers only the confirming examples of the parent's malevolence and forgets dissonant instances of the parent's good qualities. Over time, as the story hardens, it becomes more difficult to see the whole parent — the mixture of good and bad, strengths and flaws, good intentions and unfortunate blunders.

Memories create our stories, but our stories also create our memories. Once we have a narrative, we shape our memories to fit into it. In a series of experiments, Barbara Tversky and Elizabeth Marsh showed how we "spin the stories of our lives." In one, people read a story about two roommates, both of whom did something annoying and something sociable. Then they were asked to write a letter about one of the roommates, either a letter of complaint to a housing authority or a letter of recommendation to a social club. As they wrote, the study participants added elaborations and details to their letters that had not been part of the original story; if they were writing a recommendation, they might add, "Rachel is bubbly." Later, when they were asked to recall the original story as accurately as possible, their memories had become biased in the direction of the letter they had written.[8] They remembered the false details they had added and forgot the dissonant information they had not written about.

To show how memory changes to fit our stories, psychologists study how memories evolve over time: If your memories of the same people change, becoming positive or negative depending on what is happening in your life now, then it's all about you, not them. This process happens so gradually that it can be a jolt to realize you ever felt differently. "A few years back I found a diary that I wrote as a teen," a woman wrote to the advice columnist Dear Amy. "It was filled with insecurity and anger. I was shocked to read that I had ever felt that way. I consider my relationship with my mom to be very close, and I don't remember any major problems, though the diary would suggest otherwise."

The reason this letter writer doesn't "remember any major problems" was identified in two experiments by Brooke Feeney and Jude Cassidy, who showed how teenagers (mis)remember quarrels with each of their parents. Adolescents and their parents came into the lab and filled out forms listing typical topics of disagreement — personal appearance, curfews, fighting with siblings, the usual. Next, each adolescent had a ten-minute session with each parent separately to discuss and try to resolve their greatest areas of disagreement. Finally, the teenagers rated how they felt about the conflict, how intense their emotions were, what their attitudes toward their parents were, and so on. Six weeks later, they were asked to recall and rate again the conflict and their reactions to it. The teenagers who at that moment felt close to their parents remembered the quarrel as having been less intense and conflicted than they reported at the time. The teenagers who were feeling ambivalent and remote from their parents remembered the conflict as having been angrier and more bitter than they rated it at the time.[9]

Just as our current feelings about our parents shape our memories of how they treated us, our current self-concepts affect memories of our own lives. In 1962, Daniel Offer, a young resident in psychiatry, and his colleagues interviewed seventy-three teenage boys about their home lives, sexuality, religion, parents, parental discipline, and other emotionally charged topics. Offer and his colleagues were able to re-interview almost all these fellows thirty-four years later, when they were forty-eight years old, to ask them what they remembered. "Remarkably," the researchers concluded, "the men's ability to guess what they had said about themselves in adolescence was no better than chance." Most of those who remembered themselves as having been bold, outgoing teenagers had, at age fourteen, described themselves as shy. Having lived through the sexual revolution of the 1970s and 1980s, the men recalled themselves as being much more liberal and adventurous sexually than they had really been. Nearly half remembered that as teenagers, they believed that having sexual intercourse as high-school students was okay, but only 15 percent of them actually felt that way when they were fourteen. The men's current self-concepts blurred their memories, bringing their past selves into harmony with their present ones.[10]

Memories are distorted in a self-enhancing direction in all sorts of ways. Men and women alike remember having fewer sexual partners than they've actually had; they remember having far more sex with those partners than they actually had; and they remember using condoms more often than they actually did. People also remember voting in elections they didn't vote in; they remember voting for the winning candidate rather than the politician they did vote for; they remember giving more to charity than they really did; they remember that their children walked and talked at an earlier age than they really did . . . you get the idea.[11]

An appreciation of how memory works and why it so often makes mistakes can help us better evaluate many cases of he said/she said conflicts on college campuses and in news stories. We are not referring to encounters that are unambiguously coerced but to the vast majority that occur in a gray zone of human interaction. The public's typical impulse is to take one party's side and conclude that the other side is lying. But an understanding of memory and self-justification leads us to a more nuanced perspective: *a person doesn't have to be lying to be wrong.*

Sexual miscommunications, abuse, and harassment can occur

in all couples, of course — gay, straight, bi, trans — with plenty of
he said/he said and she said/she said disputes. But straight couples often struggle with an additional layer of misunderstand-
ing caused by different gender rules, norms, and expectations.
Sex researchers repeatedly find that many people rarely say what
they mean at the start of a sexual encounter, and they often don't
mean what they say. They find it difficult to say what they dislike because they don't want to hurt the other person's feelings.
They may think they want intercourse and then change their
minds. They may think they *don't* want intercourse and change
their minds. They are, in short, engaging in what social psychologist Deborah Davis calls a "dance of ambiguity." As sexologists
know from research and clinical experience, most straight men
and women, even long-term couples, communicate their sexual
wishes — including a wish not to have sex — indirectly and am-
biguously, through hints, body language, eye contact, "testing
the waters," and mind reading (which is about as accurate as . . .
mind reading). This dance of ambiguity benefits both partners;
through vagueness and indirection, each party's ego is protected
in case the other says no. Indirection saves a lot of hurt feelings,
but it also causes problems: the woman really thinks the man
should have known she wanted him to stop, and he really thinks
she gave consent.

   Davis and her colleagues Guillermo Villalobos and Richard
Leo have suggested that the primary reason for many he said/
she said disagreements is not that one side is making up an allegation or lying about a denial. Rather, each partner is providing "honest but false testimony" about what happened between
them.[12] Both parties believe they are telling the truth, but one

or both may be wrong because of the unreliability of memory — which is reconstructive in nature and exquisitely susceptible to suggestion — and because both are motivated to justify their actions. Self-justification causes individuals to distort or rewrite their memories to conform to their views of themselves, which is why they can "remember" saying things that they only thought about saying or intended to say at the time. As a result, the woman might falsely remember saying things that she thought about saying but did not say to stop the situation, because she sees herself as an assertive person who would stand up for herself. The man might falsely remember that he tried to verify the woman's consent (which he did not do), because he sees himself as a decent guy who would never rape a woman. She's not necessarily lying; she's misremembering. He's not necessarily lying; he's self-justifying.

Add alcohol to this situation, and you've got a bonfire. By far, the most well-traveled path from uncomfortable or ambiguous sexual negotiations to honest false testimony is alcohol — especially alcohol in the amounts that make participants blind drunk or cause blackouts, an epidemic problem on college campuses. Alcohol not only reduces inhibitions but also significantly impairs the cognitive interpretation of another person's behavior; men who are drunk are less likely to interpret non-consent messages accurately, and women who are drunk convey less emphatic signs of refusal. Most of all, alcohol severely impairs both partners' memory of what transpired between them. And as they form their memories, self-justification will freeze them in amber.

· · ·

If a memory is a central part of your identity, a self-serving distortion is even more likely. Ralph Haber, a distinguished cognitive psychologist, likes to tell the story of how he chose to go to graduate school at Stanford over his mother's objections. She wanted him to continue his education at the University of Michigan, he remembers, where he would be close to home, but he wanted to get far away and become more independent. "My memory has always been that when Stanford offered me admission and a fellowship, I leapt for joy, accepted with enthusiasm, and prepared to head west. A done deal!" Twenty-five years later, when Haber went back to Michigan for his mother's eightieth birthday, she handed him a shoebox of letters they had written to each other over the years. In the very first letters he pulled out, he learned that he had clearly decided to stay at Michigan and reject all his other offers. "It was my mother," he told us, "who pleaded passionately for me to change my mind" and leave. "I must have rewritten the entire history of this conflicted choice so my memory came out consistent," Haber now says, "consistent with what I actually did in leaving the shelter of home; consistent with how I wanted to see myself — being able to leave home; and consistent with my need for a loving mother who wanted me nearby." Haber's professional specialty, by the way, is autobiographical memory.

In Ralph Haber's case, the distortions of memory preserved his self-concept of always having been an independent spirit. But for most people, the self-concept is based on a belief in change, improvement, and growth. For some of us, it's based on a belief that we have changed completely; indeed, the past self seems like an entirely different person. When people have had a religious conversion, survived a disaster, suffered through cancer, or recovered from an addiction, they often feel transformed; the former self is "not me." For people who have experienced such transformations, memory helps resolve the inconsistency between their past and current selves by literally changing their perspectives. When people recall actions that are dissonant with their current view of themselves — for example, when religious people are asked to remember times they did not attend religious services when they felt they should have, or when antireligious people remember attending services — they visualize the memory from a third-person perspective, as if they were impartial observers. But when they remember actions that are consonant with their current identities, they tell a first-person story, as if they were looking at their former selves through their own eyes.[13]

What happens, though, if we think we have improved but actually haven't changed at all? Again, memory to the rescue. In one experiment, Michael Conway and Michael Ross had 106 undergraduates take a study-skills improvement program that, like many such programs, promised more than it delivered. At the start, the students rated their study skills and then were randomly assigned either to the course or to the waiting list. The training had absolutely no effect on study habits or grades. How, then, did the students who took the course justify the waste of time and effort? Three weeks later, when asked to recall as accurately as possible their own initial skills evaluation, they misremembered their skills as being far worse than they had stated at the outset, which allowed them to believe they had improved when they actually had not changed at all. Six months later, when asked to recall their grades in that course, they misremembered that too, believing their grades to have been higher than they were. The students who stayed on the waiting list for the skills program, having expended no effort, energy, or time, felt no cognitive dissonance and had nothing to justify. Having no need to distort their memories, they remembered their abilities and recent grades accurately.[14]

Conway and Ross referred to this self-serving memory distortion as "getting what you want by revising what you had." On the larger stage of life, many of us do just that: We misremember our history as being worse than it was, thus distorting our perception of how much we have improved so that we'll feel better about ourselves now.[15] All of us do grow and mature, but generally not as much as we think. This bias in memory explains why each of us feels that we have changed profoundly, but our friends, enemies, and loved ones are the same old friends, enemies, and loved ones they ever were. We run into Harry at the high-school reunion, and while Harry is busy describing how much he's learned and grown since graduation, we're nodding and saying to ourselves, "Same old Harry; a little fatter, a little balder."

The self-justifying mechanisms of memory would be just another charming and often exasperating aspect of human nature were it not for the fact that we live our lives, make decisions about people, form guiding philosophies, and construct entire narratives on the basis of memories that are often dead wrong. It's frustrating enough that things happened that we don't remember; it is scary when we remember things that never happened. Many of our mistaken memories are benign, on the level of who read us *The Wonderful O,* but sometimes they have more profound consequences, not only for ourselves but for our families, our friends, and society at large.

True Stories of False Memories

In Germany in 1995, Binjamin Wilkomirski published *Fragments,* a memoir of his horrifying childhood experiences in the concentration camps of Majdanek and Birkenau. An account of a small child's observations of Nazi atrocities and his eventual rescue and move to Switzerland, *Fragments* received extravagant praise. Reviewers compared it to the works of Primo Levi and Anne Frank. The *New York Times* said the book was "stunning" and the *Los Angeles Times* called it a "classic first-hand account of the Holocaust." In the United States, *Fragments* received the 1996 National Jewish Book Award for autobiography and memoir, and the American Orthopsychiatric Association gave Wilkomirski its Hayman Award for Holocaust and genocide study. In Britain, the book won the *Jewish Quarterly* Literary Prize; in France, it won the Prix Mémoire de la Shoah. The U.S. Holocaust Memorial Museum in Washington sent Wilkomirski on a six-city fundraising tour.

It turned out that *Fragments* was a confabulation from start to finish. Its author, whose real name was Bruno Grosjean, was not Jewish and had no Jewish ancestry. He was a Swiss musician who had been born in 1941 to an unmarried woman named Yvonne Grosjean and who was adopted several years later by a childless Swiss couple, the Dössekkers. Nor had he ever set foot in a concentration camp. His story was drawn from history books he had read, films he had seen, and Jerzy Kosinski's *The Painted Bird,* a surrealistic novel about a boy's brutal treatment during the Holocaust.[16] (Ironically, Kosinski's claim that his novel was autobiographical was later revealed to be fraudulent.)

Let's shift from Switzerland to a wealthy suburb of Boston, where Will Andrews lives. (This was the name given him by the psychologist who interviewed him.) Will is a handsome, articulate man in his forties, happily married. Will believes that he was abducted by aliens, and he has vivid memories of having been experimented on medically, psychologically, and sexually for at least ten years. In fact, he says, his alien guide became pregnant by him, producing twin boys, now eight years old, whom, he says sadly, he will never see but who play a large emotional role in his life. The abductions, he said, were terrifying and painful, but overall he is happy that he was "chosen."[17]

Are these two men guilty of fraud? Did Bruno/Binjamin Grosjean/Dössekker/Wilkomirski make up his story to become world famous, and did Will Andrews concoct memories of having been abducted by aliens to get on national talk shows? We don't think so, and we don't think that they were lying either, any more than Tom Brokaw was lying. Well, then, are these men mentally ill? Not at all. They have led perfectly reasonable lives, functioning normally, holding good jobs, having relationships, paying their bills. In fact, they are representative of the many thousands of people who have come to remember accounts of terrible suffering in their childhoods or adulthoods — experiences that were later proved beyond reasonable doubt to never have happened to them. Psychologists who have tested many of these individuals report that they do not suffer from schizophrenia or other psychotic disorders. Their mental problems, if they have any, fall within the usual range of human miseries: depression, anxiety, eating disorders, loneliness, or existential anomie.

So, no, Wilkomirski and Andrews are not crazy or deceitful, but their memories are false, and false for particular, self-justifying reasons. Their stories, so different on the face of it, are linked by common psychological and neurological mechanisms that can create false memories that nonetheless feel vividly, emotionally real. These memories do not develop overnight, in a blinding flash. They take months, sometimes years, to develop, and the stages by which they emerge are now well known to psychological scientists.

According to the Swiss historian Stefan Maechler, who interviewed Wilkomirski, his friends, his relatives, his ex-wife, and just about everyone else connected with the story, Bruno Grosjean's motivation was not calculated self-interest but self-persuasion. Grosjean spent more than twenty years transforming himself into Wilkomirski; writing *Fragments* was the last step of his metamorphosis into a new identity, not the first step of a calculated lie. "Videotapes and eyewitness reports of Wilkomirski's presentations give the impression of a man made euphoric by his own narrative," Maechler wrote. "He truly blossomed in his role as concentration-camp victim, for it was in it that he finally found himself."[18] Wilkomirski's new identity as a survivor of the Holocaust gave him a powerful sense of meaning and purpose, along with the adoration and support of countless others. How else was he going to get medals and speaking invitations? Not as a second-rate clarinetist.

Binjamin Wilkomirski, a.k.a. Bruno Grosjean, spent his first four years being bounced from place to place. His mother saw him only intermittently and finally abandoned him completely, placing him in a children's home where he lived until he was adopted by the Dössekkers. In adulthood, Wilkomirski decided that his early years were the source of his present problems, and perhaps they were. Apparently, however, an all-too-common story — being born to a single mother who couldn't care for him and eventually adopted by a kindly but formal couple — couldn't explain his difficulties dramatically enough. But what if he had not been adopted but rescued after the war and exchanged for a child named Bruno Grosjean in the orphanage? "Why else," his biographer says that Wilkomirski felt, "would he have the panic attacks that suddenly overwhelm him? Or the misshapen bump at the back of his head and the scar on his forehead? Or the nightmares that constantly plague him?"[19]

Why else? Panic attacks are a normal response to stress by those vulnerable to them. Just about everyone has bumps and scars of one kind or another; in fact, Wilkomirski's own son has the same misshapen bump in the same place, suggesting a genetic answer to that mystery. Nightmares are common in the general population, and, surprisingly, they do not necessarily reflect actual experiences. Many traumatized adults and children do not have nightmares, and many nontraumatized people do.

But Wilkomirski was not interested in these explanations. On a quest for meaning in his life, he stepped off his pyramid by deciding he would find the true reason for his symptoms in his first four lost years. At first, he didn't remember any early traumatic experiences, and the more he obsessed about his memories, the more elusive his early years felt. He started reading about the Holocaust, including survivors' accounts. He began to identify with Jews, putting a mezuzah on his door and wearing a Star of David. At the age of thirty-eight, he met Elitsur Bernstein, an Israeli psychologist living in Zurich, who would become his closest friend and adviser on his journeys into his past.

Hunting down his memories, Wilkomirski traveled to Majdanek with a group of friends, including the Bernsteins. When they arrived, Wilkomirski wept: "This was my home! This was where the children were quarantined!" The group visited the historians at the camp's archive, but when Wilkomirski asked them about the children's quarantine, they laughed at him. Very young children died or were killed, they said; the Nazis didn't run a nursery for them in a special barracks. By this time, however, Wilkomirski was too far along on his identity quest to turn back because of evidence that he was wrong, so his reaction was to reduce dissonance by dismissing the historians: "They made me look really stupid. It was a very rotten thing to do," he told Maechler. "From that moment on, I knew that I could depend more on my memory than on what is said by the so-called historians, who never gave a thought to children in their research."[20]

The next step for Wilkomirski was to go into therapy to get help for his nightmares, fearfulness, and panic attacks. He found a psychodynamically oriented analyst, Monika Matta, who analyzed his dreams and worked with nonverbal techniques, such as drawing and other methods of increasing "awareness of the body's emotions." Matta urged him to write down his memories. For people who have always remembered a traumatic or secret experience, writing can indeed be beneficial, often enabling sufferers to see their experience in a new light and begin to put it behind them.[21] But for those who are trying to remember something that never happened, writing, analyzing dreams, and drawing pictures — techniques that are the staples of many psychotherapists — are all methods that quickly conflate imagination with reality.

Elizabeth Loftus, a leading scientist in the field of memory, calls this process "imagination inflation," because the more you imagine something, the more confident you become that it really happened — and the more likely you are to inflate it into an actual memory, adding details as you go.[22] (Scientists have even tracked imagination inflation into the brain, using functional MRI to show how it works at a neural level.[23]) Giuliana Mazzoni and her colleagues asked their study participants to tell them a dream, and in return gave them a (false) "personalized" dream analysis. They told half the participants the dream meant that they had been harassed by a bully before the age of three, been lost in a public place, or been through a similar upsetting early event. Compared with control subjects, who were given no such interpretations, the dream subjects were more likely to come to believe the dream explanation had really occurred, and about half of them eventually produced detailed memories of the experience. In another experiment, people were asked to remember when their school nurse took a skin sample from their little finger to carry out a national health test. (No such test existed.) Simply imagining this unlikely scenario caused the participants to become more confident that it had happened to them. And the more confident they became, the more sensory details they added to their false memories ("the place smelled horrible").[24] Researchers have also created imagination inflation indirectly, merely by asking people to explain how an unlikely event *might* have happened. Cognitive psychologist Maryanne Garry finds that as people tell you how an event might have happened, it starts to feel real to them. Children are especially vulnerable to this suggestion.[25]

Writing turns a fleeting thought into a fact of history, and for Wilkomirski, writing down his memories confirmed his memories. "My illness showed me that it was time for me to write it all down for myself," said Wilkomirski, "just as it was held in my memory, to trace every hint all the way back."[26] Just as he rejected the historians at Majdanek who challenged his recall, he rejected the scientists who told him memory doesn't work that way.

While *Fragments* was in production, the publisher received a letter from a man alleging that Wilkomirski's story was untrue. The publisher, alarmed, contacted Wilkomirski for confirmation. Elitsur Bernstein and Monika Matta sent letters of support. "In reading Bruno's manuscript I never had any doubt as to its so-called 'authenticity,'" Bernstein wrote to the publisher. "I shall take the liberty of saying that in my judgment only someone who has experienced such things can write about them in such a way." Monika Matta, doing a little self-justification dance of her own, likewise had no doubts about the authenticity of Wilkomirski's memories or identity. Wilkomirski, she wrote, was a gifted, honest man who had "an extraordinarily precisely functioning memory" and had been profoundly shaped by his childhood experience. She wrote that she hoped that any "absurd doubts can be dispelled," because the publication of the book was so important for Wilkomirski's mental health. It was her wish, she wrote, that fate not overtake him in such a perfidious way, *"demonstrating to him yet again that he is a 'nobody.'"*[27] The publisher, convinced by the testimonials and reassurances of the experts, brought the book out on schedule. The "nobody" was somebody at last.

• • •

One evening while riding his bike across rural Nebraska, Michael Shermer was abducted by aliens. A large spaceship landed, forcing Shermer to the side of the road. Aliens descended from the ship and abducted him for ninety minutes, after which he had no memory of what had happened. Shermer's experience was not unusual; millions of Americans believe they have had some kind of encounter with UFOs or aliens. For some, it happens while they are driving long, boring miles with little change of scenery, usually at night; they gray out, losing track of time and distance, and then wonder what happened during the minutes or hours they were out of it. Some people, professional pilots among them, see mysterious lights hovering in the sky. For most, the experience occurs in the weird mental haze between sleeping and waking when they see ghosts, aliens, shadows, or spirits on their bed. Often they feel physically paralyzed, unable to move.

The bicycle racer, the driver, and the sleeper are at the top of the pyramid: Something inexplicable and alarming has happened, but what? You can live with not knowing why you woke up in a grumpy mood today, but you can't live with not knowing why you woke up with a goblin sitting on your bed. If you are a scientist or another stripe of skeptic, you will make some inquiries and learn there is a reassuring explanation for this frightening event: During the deepest stage of sleep, when dreaming is most likely to occur, part of the brain shuts down body movements so you won't go hurling yourself around the bed as you dream of chasing tigers. If you awaken from this stage before your body does, you will actually be momentarily paralyzed; if your brain is still generating dream images, you will, for a few seconds, have a waking dream. That's why those figures on the bed are dream-like, nightmarish — you *are* dreaming, but with your eyes open. Sleep paralysis, says Richard J. McNally, a Harvard psychological scientist and clinician who studies memory and trauma, is "no more pathological than a hiccup." It is quite common, he says, "especially for people whose sleep patterns have been disrupted by jet lag, shift work, or fatigue." About 30 percent of the population has had the sensation of sleep paralysis, but only about 5 percent have had the waking hallucinations as well. Just about everyone who has experienced sleep paralysis plus waking dreams reports that the feeling this combination evokes is terror.[28] It is, dare we say, an alien sensation.

Michael Shermer, a skeptic by disposition and profession, understood almost immediately what had happened to him: "My abduction experience was triggered by extreme sleep deprivation and physical exhaustion," he later wrote.[29] "I had just ridden a bicycle 83 straight hours and 1,259 miles in the opening days of the 3,100-mile nonstop transcontinental Race Across America. I was sleepily weaving down the road when my support motor home flashed its high beams and pulled alongside, and my crew entreated me to take a sleep break. At that moment a distant memory of the 1960s television series *The Invaders* was inculcated into my waking dream . . . Suddenly the members of my support team were transmogrified into aliens."

People like Shermer react to this otherworldly experience by saying, in effect, "My, what a weird and scary waking dream; isn't the brain fascinating?" But Will Andrews and the more than three million other Americans who believe they have had some kind of encounter with extraterrestrials step off the pyramid in a different direction. Clinical psychologist Susan Clancy, who interviewed hundreds of believers, found that the process moves along steadily as the possibility of alien abduction comes to seem more and more believable. "All of the subjects I interviewed," she writes, "followed the same trajectory: once they started to suspect they'd been abducted by aliens, there was no going back . . . Once the seed of belief was planted, once alien abduction was even suspected, the abductees began to search for confirmatory evidence. And once the search had begun, the evidence almost always turned up."[30]

The trigger is the frightening experience. "I woke up in the middle of the night and couldn't move," said one of her interviewees. "I was filled with terror and thought there was an intruder in the house. I wanted to scream, but I couldn't get any sound to come out. The whole thing lasted only an instant, but that was enough for me to be afraid to go back to sleep." Understandably, the person wants to make sense of what happened and looks for an explanation that might also account for other ongoing problems. "I've been depressed since as long as I can remember," said one of the people in Clancy's study. "Something is seriously wrong with me, and I want to know what it is." Others reported sexual dysfunctions, battles with weight, and odd experiences or symptoms that baffled and worried them: "I wondered why my pajamas were on the floor when I woke up"; "I've been having so many nosebleeds — I never have nosebleeds"; "I wondered where I got these coin-shaped bruises on my back."[31]

Why do these people choose an alien abduction to explain these symptoms and concerns? Why don't they consider more reasonable explanations, such as "Because I was hot in the middle of the night and took off my PJs" or "Maybe these nosebleeds are from the awful dryness in this room—I better get a humidifier" or "Maybe it's time for me to take better care of myself"? Given all the available explanations for sleep problems, depression, sexual dysfunction, and routine physical symptoms, Clancy wondered why anyone would choose the most implausible one, claiming to remember events that most of us consider impossible. The answers lie partly in American culture and partly in the needs and personalities of the *experiencers,* the term that many who believe they have been abducted use for themselves.

Experiencers come to believe that alien abduction is a reasonable explanation for their symptoms first by reading stories about it and hearing testimonials from believers. When a story is repeated often enough, it becomes so familiar that it chips away at a person's initial skepticism, even a story as unlikely as persuading people that they witnessed a demonic possession when they were children.[32] For years, the alien-abduction story was ubiquitous in American popular culture: in books, in movies, on television, on talk shows. In turn, the story fit the needs of the experiencers. Clancy found that most grew up with traditional religious beliefs, eventually rejecting them and replacing them with a New Age emphasis on channeling and alternative healing practices. This makes them more prone to fantasy and suggestion than other people, and they have more trouble with source confusion, tending to conflate things that they have thought about or experienced directly with stories they've read or heard on television. (Shermer, in contrast, recognized his aliens as coming from a 1960s television series.) Perhaps most important, the abduction explanation captures the emotional intensity and dramatic importance of the experiencers' frightening waking dreams. That explanation feels real to them, Clancy says, in a way that mundane old sleep paralysis doesn't.

The "eureka!" that experiencers feel at the fit between the alien-abduction explanation and their symptoms is exhilarating, as was the fit Wilkomirski found between the Holocaust-survivor explanation and his own difficulties. The abduction story helps experiencers explain their psychological distress and also avoid responsibility for their mistakes, regrets, and problems. "I couldn't be touched," one woman told Clancy, "not even by my husband, who's a kind and gentle man. Imagine being forty-five and not knowing what good sex was! Now I understand that it's related to what the beings did to me. I was a sexual experiment to them from an early age." All of Clancy's interviewees told her they felt changed because of their experiences, that they had become better people, that their lives had improved, and, most poignant, that their lives now had meaning. Will Andrews said, "I was ready to just give up. I didn't know what was wrong, but I knew something was missing. Today, things are different. I feel great. I know there's something out there — much bigger, more important than we are — and for some reason they chose to make their presence known to me. I have a connection with them . . . The beings are learning from us and us from them and ultimately a new world is being created. And I'll have a part in it, either directly or through the twins." Will's wife (the one on this planet) gave us an additional motive for Will's invention of invisible alien progeny when she plaintively wondered to Clancy, "Would things have been different if we had been able to have kids?"[33]

At the final stage, once the experiencers have accepted the alien-abduction explanation of their problems and retrieved their memories, they seek out other people like them and read only accounts that confirm their new explanation. They firmly reject any dissonance-creating evidence or any other way of understanding what happened to them. One of Clancy's interviewees said, "I swear to God, if someone brings up sleep paralysis to me one more time I'm going to puke. There was something in the room that night! I was spinning . . . I wasn't sleeping. I was taken."[34] Every one of the people Clancy interviewed was aware of the scientific explanation and had angrily rejected it. In Boston years ago, a debate was held between McNally and John Mack, a psychiatrist who had accepted the abductees' stories as true.[35] Mack brought an experiencer with him. The woman listened to the debate, including McNally's evidence about how people who believe they were abducted are fantasy-prone and have come to misinterpret a common sleep experience as one of seeing aliens. During the ensuing discussion, the woman said to McNally, "Don't you see, I wouldn't believe I'd been abducted if someone could just give me one reasonable alternative explanation." McNally said, "We just did."

By the end of this process, standing at the bottom of the pyramid at a far distance from skeptics like Michael Shermer, experiencers have internalized their new false memories and cannot now distinguish them from true ones. When they are brought into the laboratory and asked to describe their traumatic abductions by aliens, their heightened physiological reactions (such as heart rate and blood pressure) are as great as those of patients who suffer from posttraumatic stress disorder.[36] They have come to believe their own stories.

• • •

False memories allow people to forgive themselves and justify their mistakes, but sometimes at a high price: an inability to take responsibility for their lives. An appreciation of the distortions of memory, a realization that even deeply felt memories might be wrong, might encourage people to hold their memories more lightly, drop the certainty that their memories are always accurate, and let go of the appealing impulse to use the past to justify problems of the present. We're told to be careful what we wish for because it might come true. But we must also be careful which memories we select to justify our lives, because we will have to live by them.

Certainly one of the most powerful stories that many people wish to live by is the victim narrative. Nobody has actually been abducted by aliens (though experiencers will argue fiercely with us), but millions have survived cruelties as children: neglect, sexual abuse, parental alcoholism, violence, abandonment, the horrors of war. Many people have come forward to tell their stories: how they coped, how they endured, what they learned, how they moved on. Stories of trauma and transcendence are inspiring examples of human resilience.[37]

It is precisely because these accounts are so emotionally powerful that thousands of people have been drawn to construct me-too versions of them. A few have claimed to be Holocaust survivors, thousands have claimed to be survivors of alien abduction, and tens of thousands have claimed to be survivors of incest, rape, and other sexual traumas that allegedly were repressed from memory until they entered therapy in adulthood. Why would people claim to remember that they had suffered harrowing experiences if they hadn't, especially when that belief causes rifts with families or friends? By distorting their memories, these people can get what they want by revising what they had, and what they want is to turn their present bleak or merely mundane lives into dazzling victories over adversity. Memories of abuse also help them resolve the dissonance between "I am a smart, capable person" and "My life sure is a mess right now" with an explanation that makes them feel better about themselves and removes responsibility: "It's not my fault my life is a mess and I never became the world-class singer I could have been. Look at the horrible things my father did to me." Ellen Bass and Laura Davis made this reasoning explicit in *The Courage to Heal.* They tell readers who have no memory of childhood sexual abuse that "when you first remember your abuse or acknowledge its effects, you may feel tremendous relief. Finally there is a reason for your problems. There is someone, and something, to blame."[38]

It is no wonder, then, that most of the people who have created false memories of early suffering, like those who believe they were abducted by aliens, go to great lengths to justify and preserve their new explanations. Consider the story of a young woman named Holly Ramona, who, after a year in college, went into therapy for treatment of depression and bulimia. The therapist told her that these common problems were usually symptoms of childhood sexual abuse, which Holly denied had ever happened to her. Yet over time, at the urging of the therapist and then at the hands of a psychiatrist who administered sodium amytal (popularly but mistakenly called "truth serum"), Holly came to remember that between the ages of five and sixteen she had been repeatedly raped by her father, who even forced her to have sex with the family dog. Holly's outraged father sued both therapists for malpractice for "implanting or reinforcing false memories that [he] had molested her as a child." The jury agreed, exonerating the father and finding the therapists guilty.[39]

This ruling put Holly in a state of dissonance that she could resolve in one of two ways: She could accept the verdict, realize that her memories were false, beg her father's forgiveness, and attempt to reconcile the family that had been torn apart over her accusations. Or she could reject the verdict as a travesty of justice, become more convinced than ever that her father had abused her, and renew her commitment to recovered-memory therapy. The former, changing her mind and apologizing, would have been like turning a steamship around in a narrow river — not much room to maneuver, and hazards in every direction. The latter was by far the easier choice because of her need to justify the harm she had caused her father and the rest of her family. Much simpler to stay the course. And indeed, Holly Ramona not only vehemently rejected the verdict but went to graduate school . . . to become a psychotherapist.

· · ·

Yet once in a while someone steps forward to speak up for truth, even when the truth gets in the way of a good, self-justifying story. It's not easy, because it means taking a fresh, skeptical look at the comforting memory one has lived by, scrutinizing it from every angle for its plausibility, and, no matter how great the ensuing dissonance, letting go of it. For her entire adult life, writer Mary Karr had harbored the memory of how, as an innocent teenager, she had been abandoned by her father. That memory allowed her to feel like a heroic survivor of her father's neglect. But when she sat down to write her memoirs, she faced the realization that the story could not have been true.

"Only by studying actual events and questioning your own motives will the complex inner truths ever emerge from the darkness," she wrote.

> But how could a memoirist even begin to unearth his life's truths with fake events? At one point, I wrote a goodbye scene to show how my hard-drinking, cowboy daddy had bailed out on me when I hit puberty. When I actually searched for the teenage reminiscences to prove this, the facts told a different story: my daddy had continued to pick me up on time and make me breakfast, to invite me on hunting and fishing trips. I was the one who said no. I left him for Mexico and California with a posse of drug dealers, and then for college.
>
> This was far sadder than the cartoonish self-portrait I'd started out with. If I'd hung on to my assumptions, believing my drama came from obstacles I'd never had to overcome — a portrait of myself as scrappy survivor of unearned cruelties — I wouldn't have learned what really happened. Which is what I mean when I say God is in the truth.[40]

4

Good Intentions, Bad Science:
The Closed Loop of Clinical Judgment

It does not make any difference how beautiful your guess is. It does not make any difference how smart you are, who made the guess, or what his name is — if it disagrees with [the] experiment it is wrong. That is all there is to it.
— *Richard Feynman, physicist*

If Holly Ramona felt dissonance when the jury convicted her therapists of implanting false memories in her, how do you imagine her therapists felt? Would they be inclined to say, "Oh dear, Holly, we apologize for being so dreadfully mistaken in our treatment of your depression and eating disorders. We had better go back to school and learn a little more about memory"? The response of another psychotherapist is, we fear, more typical. A woman we will call Grace went into therapy after having a panic attack. She was not getting along with her male employer, and for the first time in her life she felt she was in a situation she could not control. But instead of treating her for panic attacks or helping her solve the job difficulty, the psychotherapist decided that Grace's symptoms meant that her father had sexually abused her when she was a child. At first, Grace embraced her therapist's interpretation; after all, the therapist was an expert on these matters. Over time, Grace, like Holly, came to believe that her father had molested her. Grace accused her father directly, cut off relations with her parents and sisters, and temporarily left her husband and son. Yet her new memories never felt right to her because they contradicted the overall history of her good and loving relationship with her father. One day she told the therapist that she no longer believed her father had abused her.

Grace's therapist might have accepted what her client told her and begun working with her on finding a better explanation for her problems. She might have read up on the latest research showing which therapeutic approach is the method of choice for panic attacks. She might have talked over the case with her colleagues to see if she was overlooking something. Grace's therapist, however, did none of these things. When Grace expressed doubts that her recovered memories were true, the therapist replied: "You're sicker than you ever were."[1]

• • •

In the 1980s and 1990s, the newly emerging evidence of the sexual abuse of children and women set off two unintended emotional epidemics, what social scientists call "moral panics." One was the phenomenon of recovered-memory therapy, in which adults went into therapy with no memory of childhood trauma and came out believing that they had been sexually molested by their parents or tortured in satanic cults, sometimes for many years, without being aware of it at the time and without any corroboration by siblings, friends, or physicians. Under hypnosis, they said, their therapists enabled them to remember the horrifying experiences they had suffered as toddlers, as infants in the crib, and sometimes even in previous lives. One woman recalled that her mother put spiders in her vagina. Another said her father had molested her from the ages of five to twenty-three, and even raped her just days before her wedding—memories she'd repressed until therapy. Others said they had been burned, although their bodies bore no scars. Some said they had been impregnated and forced to have abortions, although their bodies showed no evidence of pregnancy. Those who went to court to sue the alleged perpetrators were able to call on expert witnesses, many with impressive credentials in clinical psychology and psychiatry, to testify that these recovered memories were valid evidence of abuse.[2]

If the trauma was particularly horrible, psychiatrists claimed, the victim's personality might split into two or three or ten or a hundred personalities, causing him or her to suffer from multiple personality disorder (MPD). Before 1980, only a handful of such cases had ever been reported, and they usually claimed two personalities. (The case of "Eve" made it three faces.) Then, in 1973, *Sybil* was published. Sybil, who revealed sixteen personalities before she was through, became a national phenomenon. The book, based on an account by Sybil's psychiatrist Cornelia Wilbur, sold more than five million copies, and forty million Americans watched the 1976 two-part television special starring Joanne Woodward and Sally Field. In 1980, the American Psychiatric Association gave its official imprimatur to the syndrome by making multiple personality disorder a legitimate diagnosis, and cases began, well, multiplying. MPD clinics opened across the country to treat the growing numbers of sufferers, and by the mid-1990s there were, by various estimates, more than forty thousand people who had gone into therapy and come out believing they had dozens, even hundreds of "alters."[3]

The second moral panic arose from fears about the sexual abuse of children in daycare centers. In 1983, teachers at the McMartin Preschool in Manhattan Beach, California, were accused of committing heinous acts on the toddlers in their care, such as torturing them in satanic rituals in underground chambers, slaughtering pet rabbits in front of them, and forcing them to submit to sexual acts. Some children said the teachers had taken them flying in an airplane. The prosecution was unable to convince the jury that the children had been abused, but the case produced copycat accusations against daycare teachers across the country. Bernard "Bee" Baran, a young gay man in Massachusetts, was the first person wrongfully convicted; he spent twenty-one years in jail before a retrial freed him. The father who began the accusations against Baran had complained to the daycare center that he "didn't want no homo" teaching his child, and his wife had stated in a deposition that gays "shouldn't be allowed out in public."[4]

Baran's case was soon followed by accusations against other daycare teachers: the Little Rascals Day Care in North Carolina, Kelly Michaels in New Jersey, the Amirault family in Massachusetts, Dale Akiki in San Diego, Fran and Dan Keller in Austin, Bruce Perkins in Houston, and alleged molestation rings in Jordan, Minnesota; Wenatchee, Washington; Niles, Michigan; Miami, Florida; and dozens of other communities. The children told bizarre stories. Some said they had been attacked by a robot, molested by clowns and lobsters, or forced to eat a frog. One boy said he had been tied naked to a tree in the schoolyard in front of all the teachers and children, although no passerby noticed it and no other child verified it. Social workers and other psychotherapists were called in to assess the children's stories, do therapy with the children, and help them disclose what had happened. Many later testified in court that, in their clinical judgment, the daycare teachers were guilty.[5]

Where do epidemics go when they die? How come celebrities have not been turning up on talk shows lately to reveal their recovered memories of having been tortured as infants? Where are all the multiple personality cases? Have all the sadistic pedophiles closed down their daycare centers? Most of the teachers who were convicted have been freed on appeal, but many teachers and parents remain in prison, are confined to house arrest, or must live out their lives as registered sex offenders. Many lives were shattered and countless families have never been reunited. But cases of recovered memories of abuse in childhood still appear in the courts, in the news, in films.[6] If you look closely at these stories, many involve a therapist who helped the person "recover" his or her memories.

As for MPD, the clinics were eventually closed by successful lawsuits against the psychiatrists who had been inducing vulnerable patients to believe they had the disorder, and MPD began to fade from the cultural scene. In 2011, investigative journalist Debbie Nathan published a biography of Sybil showing that Cornelia Wilbur had virtually concocted the whole story to promote herself and sell books. Sybil did not have a childhood trauma that caused her personality to fragment; she generated her so-called personalities in response to pressures, subtle and coercive, by Wilbur, whom she wanted desperately to please — and who threatened to stop giving Sybil the drugs Wilbur had been prescribing for her and which she had become addicted to.[7]

While the epidemics have subsided, the assumptions that ignited them remain embedded in popular culture: If you were repeatedly traumatized in childhood, you probably repressed the memory of it. If you repressed the memory of it, hypnosis can retrieve it for you. If you are utterly convinced that your memories are true, they are. If you have no memories but merely suspect that you were abused, you probably were. If you have sudden flashbacks or dreams of abuse, you are uncovering a true memory. Children almost never lie about sexual matters. Watch for signs: if your child has nightmares, wets the bed, wants to sleep with a night-light, or masturbates, he or she may have been molested.

These beliefs did not pop up overnight in the cultural landscape, like mushrooms. They came from mental-health professionals who disseminated them at conferences, in clinical journals, in the media, and in bestselling books, and who promoted themselves as experts in diagnosing child sexual abuse and determining the validity of a recovered memory. Their claims were based largely on lingering Freudian (and pseudo-Freudian) ideas about repression, memory, sexual trauma, and the meaning of dreams and on their own confidence in their clinical powers of insight and diagnosis. All of the claims these therapists made have since been scientifically studied. All of them are wrong.

• • •

It is painful to admit this, but when the McMartin story first hit the news, the two of us, independently, were inclined to believe that the preschool teachers were guilty. Not knowing the details of the allegations, we mindlessly accepted the "where there's smoke, there's fire" cliché. As scientists, we should have known better; often, where there's smoke, there's just smoke. Months after the trial ended, when the full story came out — about the emotionally disturbed mother who made the first accusation and whose charges became crazier and crazier until even the prosecutors stopped paying attention to her; about how the children had been coerced over many months to "tell" by zealous social workers on a moral crusade; about how the children's stories became increasingly outlandish — we felt foolish and embarrassed that we had sacrificed our scientific skepticism on the altar of outrage. Our initial gullibility caused us plenty of dissonance, and it still does. But our dissonance is nothing compared to that of the people who were personally involved or who took a public stand, including the many psychotherapists, psychiatrists, and social workers who considered themselves skilled clinicians and advocates for children's rights.

None of us like learning that we were wrong, that our memories are distorted or confabulated, or that we made an embarrassing professional mistake. For people in any of the healing professions, the stakes are especially high. If you hold a set of beliefs that guide your practice and you learn that some of them are incorrect, you must either admit you were wrong and change your approach or reject the new evidence. If the mistakes are not too threatening to your view of your own competence and if you have not taken a public stand defending them, you will probably willingly change your approach, grateful to have a better one. But if some of those mistaken beliefs have made your clients' problems worse, torn up your clients' families, or sent innocent people to prison, then you, like Grace's therapist, will have serious dissonance to resolve.

It's the Semmelweis phenomenon that we described in the introduction. Semmelweis discovered that when his medical students washed their hands before attending women in labor, fewer women died of childbed fever. Why didn't his colleagues say, "Hey, Ignaz, thank you so much for finding the reason for the tragic, unnecessary deaths of our patients"? Before these physicians could accept his simple, lifesaving intervention, they would have had to admit that they had been the cause of the deaths of all those women in their care. This was an intolerable realization, for it went straight to the heart of the physicians' view of themselves as medical experts and wise healers. And so they told Semmelweis, in essence, to get lost and take his stupid ideas with him. Because their stubborn refusal to accept Semmelweis's evidence — the lower death rate among patients whose doctors had washed their hands — happened long before the era of malpractice suits, we can say with assurance that they were acting out of a need to protect their egos, not their income. Medicine has advanced since their day, but the need for self-justification hasn't budged.

Most occupations are ultimately, if slowly, self-improving and self-correcting. If you are a physician today, you wash your hands and you wear gloves, and if you forget, your colleagues, nurses, or patients will remind you. If you run a toy company and make a mistake in predicting that your new doll will outsell Barbie, the market will let you know. If you are a scientist who faked the data on your cloned sheep and then tried to pull the wool over your colleagues' eyes, the first lab that cannot replicate your results will race to tell the world you're a fraud. If you are an experimental psychologist and make a mistake in the design of your experiment or in your analysis of the results, your colleagues and critics will be eager to inform you, the rest of the scientific community, and everyone on the ex-planet Pluto. Naturally, not all scientists are scientific — that is, open-minded and willing to give up their strong convictions or admit that conflicts of interest might taint their research. But even when an individual scientist is not self-correcting, science eventually is.

The mental-health professions are different. Professionals in these fields have an amalgam of credentials, training, and approaches that often bear little connection to one another. Imagine that the profession of law consisted of people who attended law school, studied each area of the law diligently, and passed the grueling bar exam as well as people who did nothing but pay seventy-eight dollars for a weekend course in courtroom etiquette, and you will have a glimpse of the problem. And which lawyer would you want defending you?

In the profession of psychotherapy, clinical psychologists are the closest equivalent to traditionally trained lawyers. Most have a PhD, and if they earned the degree from a major university rather than from an independent therapy mill, they have a knowledge of basic psychological findings. Some do research themselves to determine the ingredients of successful therapy or the origins of emotional disorders. But whether or not they personally do research, they tend to be well versed in psychological science and know which kind of therapy is demonstrably most effective for what problem. They would know that cognitive and behavioral methods are the psychological treatments of choice for panic attacks, depression, eating disorders, insomnia, chronic anger, and other emotional disorders. These methods are often as effective or more effective than medication.[8]

In contrast, most psychiatrists, who have medical degrees, learn about medicine and medication, but they rarely learn much about basic research in psychology. Throughout the twentieth century, they were generally practitioners of Freudian psychoanalysis or one of its offshoots; you needed an MD to be admitted to a psychoanalytic training institute. As the popularity of psychoanalysis declined and the biomedical model of disorder gained the upper hand, most psychiatrists began treating patients with medication rather than any form of talk therapy. Yet while psychiatrists learn about the brain, many still learn almost nothing about nonmedical causes of emotional disorders or about the questioning, skeptical essence of science. Anthropologist Tanya Luhrmann spent four years studying residents in psychiatry, attending their classes and conventions, observing them in clinics and emergency rooms. She found that residents were not expected to read much; they were expected to absorb the lessons handed them without debate or question. The lectures they attended offer practical skills, not intellectual substance; a lecturer would talk about what to do in therapy rather than why the therapy helps or what kind of therapy might be best for a given problem.[9]

Finally, there are the many people who practice the many different forms of psychotherapy. Some have a master's degree in psychology, counseling, or clinical social work and are licensed

in a specialty, such as marriage and family therapy. Some, however, have no training in psychology at all; some don't even have
a college degree. The word *psychotherapist* is largely unregulated;
in many states, anyone can say that he or she is a therapist without having any training in anything.

In the past decades, as the number of mental-health practitioners of all kinds has soared, most counseling-psychology and
psychotherapy-training programs have cut themselves off from
their scientifically trained cousins in university departments of
psychology.[10] "What do I need to know statistics and research
for?" many graduates of these programs ask. "All I need to know
is how to do therapy, and for that, I mostly need clinical experience." In some respects, they are right. Therapists are constantly
making decisions about the course of treatment: What might be
beneficial now? What direction should we go? Is this the right
time to risk challenging my client's story or will I challenge him
right out of the room? Making these decisions requires experience with the infinite assortment of quirks and passions of the
human psyche, that heart of darkness and love.

Moreover, by its very nature, psychotherapy is a private interaction between the therapist and the client. No one is looking
over the therapist's shoulder in the intimacy of the consulting
room, eager to pounce if he or she does something wrong. Yet
the inherent privacy of the interaction means that therapists
who lack training in science and skepticism have no internal corrections to the self-protecting cognitive biases that afflict us all.
What these therapists see confirms what they believe, and what
they believe shapes what they see. It's a closed loop. Did my client improve? Excellent; what I did was effective. Did my client

remain unchanged or get worse? That's unfortunate, but she *is* resistant to therapy and deeply troubled; besides, sometimes the client has to get worse before she can get better. Do I believe that repressed rage causes sexual difficulties? If so, my client's erection problem must reflect his repressed rage at his mother or his wife. Do I believe that sexual abuse causes eating disorders? If so, my client's bulimia must mean she was molested as a child.

We want to be clear that some clients *are* resistant to therapy and *are* deeply troubled. This chapter is not an indictment of therapy, any more than pointing out the mistakes of memory means that all memory is unreliable or that the conflicts of interest among scientists means that all scientists do tainted research. Our intention is to examine the kinds of mistakes that can result from the closed loop of clinical practice and show how self-justification perpetuates them.

For anyone in private practice, skepticism and science are ways out of the closed loop. Skepticism teaches therapists to be cautious about taking what their clients tell them at face value. If a woman says her mother put spiders in her vagina when she was three, the skeptical therapist can feel empathy without believing that this event literally happened. If a child says his teachers took him flying in a plane full of clowns and frogs, the skeptical therapist might be charmed by the story without believing that teachers actually chartered a private jet (on their salary, no less). Scientific research provides therapists with ways of improving their clinical practice and of avoiding mistakes. If you are going to use hypnosis, you had better know that while hypnosis can help clients learn to relax, manage pain, and quit smoking, you should never use it to help your client retrieve memories, because your willing, suggestible client will often make up a memory that is unreliable.[11]

Yet today there are many thousands of psychiatrists, social workers, counselors, and psychotherapists who go into private practice with neither skepticism nor evidence to guide them. Paul Meehl, who achieved great distinction as both a clinician and a scientific researcher, observed that when he was a student, the common factor in the training of all psychologists was "the general scientific commitment not to be fooled and not to fool anyone else. Some things have happened in the world of clinical practice that worry me in this respect. That skepsis, that passion not to be fooled and not to fool anyone else, does not seem to be as fundamental a part of all psychologists' mental equipment as it was a half century ago . . . I have heard of some psychological testimony in courtrooms locally in which this critical mentality appears to be largely absent."[12]

An example of the problem Meehl feared can be seen in the deposition of a prominent psychiatrist, Bessel van der Kolk, who testified frequently on behalf of plaintiffs in repressed-memory cases. Van der Kolk explained that as a psychiatrist, he had had medical training and a psychiatric residency, but he had never taken a course in experimental psychology.

Q: Are you aware of any research on the reliability or the validity of clinical judgment or clinical predictions based on interview information?

A: No.

Q: What's your understanding of the current term "disconfirming evidence"?

A : I guess that means evidence that disconfirms treasured notions that people have.

Q : What's the most powerful piece of disconfirming evidence
that you're aware of for the theory that people can repress
memories or that they can block out of their awareness a series of traumatic events, store those in their memory, and recover those with some accuracy years later?

A : What's the strongest thing against that?

Q : Yes. What's the strongest piece of disconfirming evidence?

A : I really can't think of any good evidence against that . . .

Q : Have you read any literature on the concept of false memories using hypnosis?

A : No.

Q : Is there research on whether clinicians over a period of years
develop more accurate clinical judgment?

A : I don't know if there is, actually . . .

Q : Is [there] a technique that you use to distinguish true and
false memories?

A : We all, we all as human beings are continuously faced with
whether we believe what somebody feeds us or not, and we
all make judgments all the time. And there is such a thing as
internal consistency, and if people tell you something with
internal consistency and with appropriate affect, you tend to
believe that the stories are true.[13]

At the time of this deposition, van der Kolk had not read
any of the voluminous research literature on false memories or
how hypnosis can create them, nor was he aware of the docu-

mented unreliability of "clinical predictions based on interview information." He had not read any of the research disconfirming his belief that traumatic memories are commonly repressed. Yet he testified frequently and confidently on behalf of plaintiffs in repressed-memory cases. Like many clinicians, he is confident that he knows when a client is telling the truth, whether a memory is true or false, based on his clinical experience; the clues are whether the client's story has "internal consistency" and whether the client recounts the memory with appropriate emotion — that is, whether the client really *feels* the memory is true. The problem with this reasoning, however, is that, as we saw in the previous chapter, thousands of mentally healthy people believe they were abducted by aliens and are able to report, with all the appropriate feeling, internally consistent stories of the bizarre experiments they believe they endured. As research psychologist John Kihlstrom observed, "The weakness of the relationship between accuracy and confidence is one of the best-documented phenomena in the 100-year history of eyewitness memory research,"[14] but van der Kolk was unaware of a finding that just about every undergraduate who has taken Psychology 101 would know.

As evidence accumulated on the fallibility of memory and the many confabulations of recovered-memory cases, the promoters of this notion did not admit error; they simply changed their view of the mechanism by which traumatic memories are allegedly lost. It's not repression at work anymore, but dissociation; the mind somehow splits off the traumatic memory and banishes it to the suburbs. This shift allowed them to keep testifying, without batting an eye or ruffling a feather, as scientific experts in cases of recovered memories.

Consider the 2014 testimony of Christine Courtois, a counseling psychologist who has been a proponent of recovered-memory therapy for more than thirty years. (Her practice was closed in 2016.) She was brought in as an expert in a civil action on behalf of a plaintiff who claimed he had been molested as a boy by the defendant but who had only recently come to remember the abuse. A pretrial hearing was held to determine whether there was a reliable scientific basis for his claim. Usually, there is a statute of limitations in civil suits, including child-abuse cases. But many courts have ruled that the clock is stopped on the time limitation if the plaintiff at first has no knowledge of the harm that he or she later comes to remember. The courts agree that a person's being in a coma stops the clock on the statute of limitations, but they do not agree about whether repressed memories also stop the clock. The resolution rests on the scientific merit of the claim that traumatic memories can be repressed or dissociated. If they can be, then civil and criminal actions can be brought within a certain time after the plaintiff *remembers* being molested rather than after the molestation itself. This is why, in such cases, the plaintiff's attorneys bring in the biggest clinical guns they can find to testify about the existence of repression — or, nowadays, dissociation. Thanks to the huge popularity of neuroscience and studies of the brain, these experts, who testified for years about the existence of repressed memories, now wave around vague references to parts of the brain to support their new belief in the existence of dissociated memories, as can be seen in Dr. Courtois's incoherent testimony:

A. Which has to do with over-inhibition of response to the stress trauma in the individual's brain, and different parts of the brain either light up or shut down and show a differential response. So the dissociative part is related to the depersonalization, derealization from the experience, and those kind of mechanisms would make it easier to sequester that information. It apparently doesn't go away and is accessible later on, but it is sequestered in the brain.

Some of the research is also showing that the brain of traumatized children, the brains of traumatized children versus children who are not traumatized by virtue of their differential experience, which often starts at a very young age, the brains are different and the brain development is different, the brain function, the brain structure is different. Which may have implications for memory retention, memory encoding, memory retrieval later on.[15]

Are you impressed? If so, you are not alone. This is the kind of language that sounds serious and scientific, but on closer inspection reveals itself to be gobbledygook. Different parts of the brain are doing what? Which parts? The brain structure is different in trauma victims? How? "Implications for memory retention" means what, exactly? "Sequestered in the brain"? Where? In a small closet off the corpus callosum? In their paper "The Seductive Appeal of Neuroscience Explanations," Deena Weisberg and her colleagues demonstrated that if you give one group of laypeople a straightforward explanation of some behavior and another group the same explanation but with vague references to the brain thrown in ("brain scans indicate" or "the frontal-lobe brain circuitry known to be involved"), people assume the latter is more scientific — and therefore more real. Many intelligent people, including psychotherapists, fall prey to the seductive appeal of this language, but laypeople aren't called upon in court to try to explain what it means.[16]

No one is suggesting that UN observers disturb the privacy of the therapeutic encounter or that all therapists should start doing their own research. An understanding of how to think scientifically may not aid therapists in the subjective process of helping a client who is searching for answers to existential questions. But it matters profoundly when therapists claim expertise and certainty in domains in which their unverified clinical opinions can ruin lives. The scientific method consists of the use of procedures designed to show not that our predictions and hypotheses are right, *but that they might be wrong*. Scientific reasoning is useful to anyone in any job because it makes us face the possibility, even the dire reality, that we were mistaken. It forces us to confront our self-justifications and put them on public display for others to puncture. At its core, therefore, science is a form of arrogance control.

The Problem of the Benevolent Dolphin

Every so often, a heartwarming news story tells of a shipwrecked sailor who was on the verge of drowning in a turbulent sea. Suddenly, a dolphin popped up at his side and, gently but firmly, nudged the swimmer safely to shore. Dolphins must really like human beings, enough to save us from drowning! But wait — are dolphins aware that humans don't swim as well as they do? Are they actually intending to be helpful? To answer that question, we would need to know how many shipwrecked sailors have been gently nudged *farther* out to sea by dolphins, there to drown and never be heard from again. We don't know about those cases, because the swimmers don't live to tell us about their evil-dolphin experiences. If we had that information, we might conclude that dolphins are neither benevolent nor evil; they are just being playful.

Sigmund Freud himself fell victim to the flawed reasoning of the benevolent-dolphin problem. When his fellow analysts questioned his notion that all men suffer from castration anxiety, he was amused. He wrote: "One hears of analysts who boast that, though they have worked for dozens of years, they have never found a sign of the existence of the castration complex. We must bow our heads in recognition of . . . [this] piece of virtuosity in the art of overlooking and mistaking."[17] So if analysts see castration anxiety in their patients, Freud was right, and if they fail to see it, they are "overlooking" it, and Freud is still right. Men themselves cannot tell you if they feel castration anxiety, because it's unconscious, but if they deny that they feel it, they are in denial.

What a terrific theory! No way for it to be wrong. But that is the very reason that Freud, for all his illuminating observations about civilization and its discontents, was not doing science. For any theory to be scientific, it must be stated in such a way that it can be shown to be false as well as true. If any outcome confirms your hypothesis that all men unconsciously suffer from castration anxiety (or that intelligent design, rather than evolution, accounts for the diversity of species, or that your favorite psychic would accurately have predicted 9/11 if only she hadn't been taking a shower that morning), your beliefs are a matter of faith, not science. Freud, however, saw himself as the consummate scientist. In 1934, the American psychologist Saul Rosenzweig wrote to him, suggesting that Freud subject his psychoanalytic assertions to experimental testing. "The wealth of dependable observations on which these assertions rest make them independent of experimental verification," Freud replied loftily. "Still, [experiments] can do no harm."[18]

Because of the confirmation bias, however, the "dependable observation" is not dependable. Clinical intuition — "I know it when I see it" — is the end of the conversation to many psychiatrists and psychotherapists but the start of the conversation to the scientist: "A good observation, but what exactly have you seen, and how do you know you are right?" Observation and intuition without independent verification are unreliable guides; like roguish locals misdirecting the tourists, they occasionally send us off in the wrong direction.

Although there are few orthodox Freudians anymore, there are many psychodynamic schools of therapy, so called because they derive from Freud's emphasis on unconscious mental dynamics. Most of these programs are unconnected to university departments of psychological science (though some are still part of training for psychiatric residents), and their students learn little to nothing about scientific methods or even about basic psychological findings. And then there are the many unlicensed therapists who don't know much about psychodynamic theories but nonetheless have uncritically absorbed the Freudian language that permeates the culture — notions of regression, denial, and repression. What unites these clinical practitioners is their misplaced reliance on their own powers of observation and the closed loop it creates. Everything they see confirms what they believe.

One danger of the closed loop is that it makes practitioners vulnerable to logical fallacies. Consider the famous syllogism "All men are mortal; Socrates is a man; therefore Socrates is mortal." So far, so good. But just because all men are mortal, it does not follow that all mortals are men, and it certainly does not follow that all men are Socrates. Yet the recovered-memory movement was based on the logical fallacy that if some women who have been sexually abused in childhood develop depression, eating disorders, and panic attacks, then all women who suffer from depression, eating disorders, and panic attacks must have been sexually abused. Accordingly, many psychodynamic clinicians began pushing their unhappy clients to rummage around in their pasts to find supporting evidence for their theory. But some clients denied that they had been abused. What to do with this dissonant response? The answer came in Freud's idea that the unconscious actively represses traumatic experiences, particularly those of a sexual nature. That explains it! That explains how Holly Ramona could forget that her father raped her for eleven years.

Once these clinicians had latched on to repression to explain why their clients were not remembering traumatic sexual abuse,

you can see why some felt justified, indeed professionally obligated, to do whatever it took to pry that repressed memory out
of there. Because the client's denials are all the more evidence of
repression, strong methods are called for. If hypnosis won't do
it, let's try sodium amytal ("truth serum"), another intervention
that merely relaxes a person and increases the chances of false
memories.[19]

Of course, many of us intentionally avoid painful memories
by distracting ourselves or trying not to think about them, and
many of us have had the experience of suddenly recalling an embarrassing memory, one we thought long gone, when we are in
a situation that evokes it. The situation provides what memory
scientists call retrieval cues, familiar signals that reawaken the
memory.[20]

Psychodynamic therapists, however, claim that repression
is entirely different from the normal mechanisms of forgetting
and recall. They think it explains why a person can forget years
and years of traumatic experiences, such as repeated rape. Yet
in his meticulous review of the experimental research and the
clinical evidence, presented in his book *Remembering Trauma,*
clinical psychologist Richard McNally concluded: "The notion
that the mind protects itself by repressing or dissociating memories of trauma, rendering them inaccessible to awareness, is a
piece of psychiatric folklore devoid of convincing empirical support."[21] Overwhelmingly, the evidence shows just the opposite.
The problem for most people who have suffered traumatic experiences is not that they forget them but that they cannot forget
them; the memories keep intruding.

Thus, people do not repress the memory of being tortured in

prison, being in combat, or being the victim of a natural disaster (unless they suffered brain damage at the time), although details of even these horrible experiences are subject to distortion over the years, as are all memories. "Truly traumatic events — terrifying, life-threatening experiences — are never forgotten, let alone if they are repeated," says McNally. "The basic principle is: if the abuse was traumatic at the time it occurred, it is unlikely to be forgotten. If it was forgotten, then it was unlikely to have been traumatic. And even if it was forgotten, there is no evidence that it was blocked, repressed, sealed behind a mental barrier, inaccessible."

This is obviously disconfirming information for clinicians committed to the belief that people who have been brutalized for years will repress the memory. If they were right, surely Holocaust survivors would be leading candidates for repression. But as far as anyone knows, and as McNally documents, no survivors of the Holocaust have forgotten or repressed what happened to them. Recovered-memory advocates have a response to that evidence too — they distort it. In one study conducted forty years after the war, survivors of Camp Erika, a Nazi concentration camp, were asked to recall what they had endured there. When their current recollections were compared with depositions they had provided when they were first released, it turned out that the survivors remembered what happened to them with remarkable accuracy. Any neutral observer would read this research and say, "How incredible! They were able to recall all those details after forty years." Yet one team of recovered-memory advocates cited this study as evidence that "amnesia for Nazi Holocaust camp experiences has also been reported." What was reported was nothing remotely like amnesia. Some survivors failed to recall a few violent events among a great many similar ones, and some had forgotten a few specifics, such as the name of a sadistic guard. This is not repression; it is the normal forgetting of details that all of us experience over the years.[22]

Clinicians who believe in repression see it everywhere, even where no one else does. But if everything you observe in your clinical experience is evidence to support your beliefs, what would you consider counterevidence? What if your client has no memory of abuse not because she is repressing, but because it never happened? What could ever break you out of the closed loop? To guard against the bias of our own direct observations, scientists invented the control group: the group that *isn't* getting the new therapeutic method, the patients who *aren't* getting the new drug. Most people understand the importance of control groups in a study of a new drug's effectiveness, because without a control group, you can't say if people's positive response is due to the drug or to the placebo effect — the general expectation that the drug will help them. One study of women who had complained of sexual problems found that 41 percent said that their libido returned when they took Viagra. So, however, did 43 percent of the control group who took a sugar pill.[23] (This study showed conclusively that the organ most responsible for sexual excitement is the brain.)

Obviously, if you are a psychotherapist, you can't randomly put some of your clients on a waiting list and give others your serious attention; the former will find another therapist pronto. But if you are not trained to be aware of the benevolent-dolphin problem and if you are absolutely, positively convinced that your views are right and your clinical skills unassailable, you can make serious errors. A clinical social worker explained why she had decided to remove a child from her mother's custody: the mother had been physically abused as a child, and "we all know," the social worker said to the judge, that meant she would almost certainly be an abusive parent herself. This assumption of the cycle of abuse came from observations of confirming cases: abusive parents, in jail or in therapy, reporting that they had been severely beaten or sexually abused by their own parents. What is missing are the *disconfirming* cases, the abused children who do not grow up to become abusive parents. They are invisible to social workers and other mental-health professionals because, by definition, they don't end up in prison or treatment. Research psychologists who have done longitudinal studies, following children over time, have found that while being physically abused as a child is associated with an increased chance of becoming an abusive parent, the great majority of abused children — nearly 70 percent — do not repeat their parents' cruelties.[24] If you are doing therapy with a victim of parental abuse or with an abusive parent, this information may not be relevant to you. But if you are in a position to make predictions that will affect whether, say, a parent should lose custody of a child, it most surely is.

Similarly, suppose you are doing therapy with children who have been sexually molested. They touch your heart, and you take careful note of their symptoms: they are fearful, wet the bed, want to sleep with a night-light, have nightmares, masturbate, or expose their genitals to other children. After a while, using those symptoms as a checklist, you will probably become pretty confident of your ability to determine whether a child has been abused. You might give a toddler an anatomically correct doll to play with on the grounds that what he or she cannot reveal in words may be revealed in play. One of your young clients pounds a stick into a doll's vagina. Another scrutinizes a doll's penis with alarming concentration for a four-year-old.

Therapists who have not been trained to think scientifically will probably not wonder about the invisible cases — the children they don't see as clients. They probably will not think to ask how common the symptoms of bed-wetting, sex play, and fearfulness are in the general population of children. When researchers did ask, they found that children who have not been molested are also likely to masturbate and be sexually curious; temperamentally fearful children are also likely to wet the bed and be scared of the dark.[25] Even children who have been molested show no predictable set of symptoms, something scientists learned only by observing children's reactions over time instead of by assessing them once or twice in a clinical interview. A review of forty-five studies that followed sexually abused children for up to eighteen months found that although these children at first had more symptoms of fearfulness and sexual acting-out than nonabused children, "no one symptom characterized a majority of sexually abused children [and] approximately one third of victims had no symptoms . . . The findings suggest the absence of any specific syndrome in sexually abused children."[26]

Moreover, children who have not been abused do not appreciably differ from abused children in how they play with anatomically detailed dolls; those prominent genitals are pretty interesting. Some children do bizarre things and it doesn't mean anything at all except that the dolls are unreliable as diagnostic tests.[27] In one study headed by two eminent developmental psychologists, Maggie Bruck and Stephen Ceci, a child pounded a stick into the doll's vagina to show her parents what supposedly had happened to her during a doctor's exam that day.[28] The (videotaped) doctor had done no such thing, but you can imagine how you would feel if you watched your daughter playing so violently with the doll, and a psychiatrist told you solemnly it meant she had been molested. You would want that doctor's hide.

Many therapists feel extremely confident of their ability to determine whether a child has been molested because, they say, they have years of clinical experience to back up their judgments. Yet study after study shows that their confidence is unwarranted. In one important study, clinical psychologist Thomas Horner and his colleagues examined the evaluations provided by a team of expert clinicians in a case in which a father was accused of molesting his three-year-old daughter. The experts reviewed transcripts, watched interviews of the child and videotapes of parent-child exchanges, and reviewed the clinical findings. They had identical information, but some were convinced the abuse had occurred while others were just as convinced it had never happened. The researchers then recruited 129 other mental-health specialists and asked them to assess the evidence in this case, estimate the likelihood that the little girl had been molested by her father, and make a recommendation regarding custody. Again, the results ranged from certainty that the child had been molested to certainty that she had not. Some wanted to forbid the father to see his daughter ever again; others wanted to give him full custody. Those experts who were prone to believe that sexual abuse is rampant in families were quick to interpret ambiguous evidence in ways that supported that belief; those who were skeptical did not. For the unskeptical experts, the researchers said, "believing is seeing."[29]

To date, hundreds of studies have demonstrated the unreliability of clinical predictions. This evidence is dissonance-creating news to the mental-health professionals whose self-confidence rests on the belief that their expert assessments are extremely accurate.[30] When we said that science is a form of arrogance control, that's what we mean.

• • •

"Believing is seeing" was the principle that created every one of the daycare scandals of the 1980s and 1990s. Just as in the McMartin case, each began with an accusation from a disturbed or homophobic parent or the whimsical comments of a child, which provoked an investigation, which provoked panic. At the Wee Care Nursery School in New Jersey, a four-year-old was having his temperature taken rectally at his doctor's office when he said, "That's what my teacher [Kelly Michaels] does to me at school."[31] The child's mother notified the state's child protection agency. The agency brought the child to a prosecutor's office and gave him an anatomical doll to play with. The boy inserted his finger into the rectum of the doll and said that two other boys had had their temperature taken that way too. Parents of children in the preschool were told to look for signs of abuse in their own children. Professionals were called in to interview the children. Before long, the children were claiming that Kelly Michaels had, among other things, licked peanut butter off their genitals, made them drink her urine and eat her feces, and raped them with knives, forks, and toys. These acts were said to have occurred during school hours over a period of seven months, although no child had complained, and parents, who could come and go as they pleased, never witnessed any abuse or noticed any problems in their children.

Kelly Michaels was convicted of 115 counts of sexual abuse and sentenced to forty-seven years in prison. She was released after five years when an appeals court ruled that the children's testimony had been tainted by how they had been interviewed. And how was that? With the confirmation bias going at full speed and no reins of scientific caution to restrain it, a deadly combination that was the hallmark of the interviews of children conducted in all the daycare cases. Here is how Susan Kelley, a pediatric nurse who interviewed children in a number of these cases, used Bert and Ernie puppets to "aid" the children's recall:

KELLEY: Would you tell Ernie?

CHILD: No.

KELLEY: Ah, come on [pleading tone]. Please tell Ernie. Please tell me. Please tell me. So we could help you. Please . . . You whisper it to Ernie . . . Did anybody ever touch you right there? [pointing to the vulva of a girl doll]

CHILD: No.

KELLEY: [pointing to the doll's posterior] Did anybody touch your bum?

CHILD: No.

KELLEY: Would you tell Bert?

CHILD: They didn't touch me!

KELLEY: Who didn't touch you?

CHILD: Not my teacher. Nobody.
KELLEY: Did any big people, any adult, touch your bum there?
CHILD: No.[32]

"Who didn't touch you?" We are entering the realm of *Catch-22,* Joseph Heller's great novel, in which the colonel with the fat mustache says to Clevinger: "What did you mean when you said we couldn't punish you?" Clevinger replies: "I didn't say you couldn't punish me, sir." Colonel: "When didn't you say that we couldn't punish you?" Clevinger: "I always didn't say that you couldn't punish me, sir."

At the time, the psychotherapists and social workers who were called on to interview children believed that molested children wouldn't tell you what happened to them unless you pressed them by persistently asking leading questions, because they were scared or ashamed. In the absence of research, this was a reasonable assumption, and clearly it is sometimes true. But when does pressing slide into coercion? Psychological scientists have conducted experiments to investigate various aspects of children's memory and testimony. How do children understand what adults ask them? Do their responses depend on their age, verbal abilities, and the kinds of questions they are asked? Under what conditions are children likely to be telling the truth, and when are they likely to be suggestible, to say that something happened when it did not?[33]

In an experiment with preschool children, Sena Garven and her colleagues used interview techniques that were based on the actual transcripts of interrogations of children in the McMartin case. A young man visited children at their preschool, read them a story, and handed out treats. He did nothing aggressive, inappropriate, or surprising. A week later an experimenter questioned the children about the man's visit. She asked one group leading questions such as "Did he shove the teacher? Did he throw a crayon at a kid who was talking?" She asked a second group the same questions but added the influence techniques the McMartin interrogators had used: she told the children what other kids had supposedly said, expressed disappointment if their answers were negative, and praised children for making allegations. The children in the first group, who got merely the leading questions, said "Yes, it happened" to about 15 percent of the false allegations about the man's visit; not a high percentage, but not a trivial one either. In the second group, however, the one in which influence tactics had been added, the three-year-olds said "Yes, it happened" to over 80 percent of the false allegations suggested to them, and the four- to six-year-olds said yes to about half the allegations. And those results occurred after interviews lasting only five to ten minutes; in actual criminal investigations, interviewers often question children repeatedly over weeks and months. In a similar study, this time with five- to seven-year-olds, investigators found they could easily influence the children to answer yes to preposterous questions, such as "Did Paco take you flying in an airplane?" What was more troubling was that within a short time, many of the children's inaccurate statements had crystallized into stable, but false, memories.[34]

Research like this has enabled psychologists to improve their methods of interviewing children. Their goal is to help children who have been abused disclose what happened to them but without increasing the suggestibility of children who have not been abused. The scientists have shown that children under age five often cannot tell the difference between something they were told and something that actually happened to them. If pre-schoolers overhear adults exchanging rumors about some event, many of the children will later come to believe they actually experienced the event themselves.[35] In all these studies, the most powerful finding is that adults are highly likely to taint an interview when they go into it already convinced that a child has been molested. When that is so, there is only one "truth" they are prepared to accept when they ask the child to tell the truth. Like Susan Kelley, they never accept the child's *no; no* means the child is denying or repressing or afraid to tell. The child can do nothing to convince the adult she has not been molested.

The adult might even be the child's parent. Twenty-one years after the McMartin trials, Kyle Zirpolo, who had been one of the children who testified against the preschool teachers, publicly apologized in the *Los Angeles Times*. He knew at the time he was lying, he said, but he did so to please his punitive, police-officer stepfather, who was convinced the daycare teachers were pedo-philes. As Zirpolo said:

> But the lying really bothered me. One particular night stands out in my mind. I was maybe 10 years old and I tried to tell my mom that nothing had happened. I lay on the bed crying hysterically — I wanted to get it off my chest, to tell her the truth. My mother kept asking me to please tell her what was the matter. I said she would never believe me. She persisted: "I promise I'll believe you! I love you so much! Tell me what's bothering you!" This

went on a long time: I told her she wouldn't believe me, and she kept assuring me she would. I remember finally telling her, "Nothing happened! Nothing ever happened to me at that school."

She didn't believe me.[36]

Zirpolo believed his mother could not accept the truth — that he was not molested — because if she did, "how [could] she explain all the family's problems?" She and his stepfather never listened to him, he said, never expressed relief that he hadn't been harmed, and never saw any of the movies or read any of the books that questioned the prosecution's handling of the case.

We can understand why so many Susan Kelleys, prosecutors, and parents have been quick to assume the worst; no one wants to let a child molester go free. But no one should want to contribute to the conviction of an innocent adult either. Today, informed by years of experimental research with children, the National Institute of Child Health and Human Development and some individual states have drafted new model protocols for social workers, police investigators, and others who conduct child interviews.[37] These protocols emphasize the hazards of the confirmation bias; they instruct interviewers to test the hypothesis of possible abuse and not assume they know what happened. The guidelines recognize that most children will readily disclose actual abuse, and some need prodding; the guidelines also caution against the use of techniques known to produce false reports.

This change, from the uncritical "believe the children" to the more discerning "understand the children," reflects a recognition that mental-health professionals need to think more like scientists and less like advocates; they must weigh all the evidence fairly and consider the possibility that their suspicions are unfounded. If they do not, it will not be justice that is served, but self-justification.

Science, Skepticism, and Self-justification

When psychiatrist Judith Herman published *Father-Daughter Incest* in 1981, the patients she described remembered what had happened to them all too clearly. At the time, feminist clinicians like Herman were working to raise public awareness of rape, child abuse, incest, and domestic violence. They were not claiming that their clients had repressed their memories; rather, these women said they had chosen to remain silent because they felt frightened, ashamed, and certain that no one would believe them. There is no entry for *repression* in the index of *Father-Daughter Incest*. Yet within ten years Herman had become a recovered-memory advocate; the first sentence of her 1992 book *Trauma and Recovery* is "The ordinary response to atrocities is to banish them from consciousness." How did Herman and other highly experienced clinicians move from believing that traumatic experiences are rarely if ever forgotten to believing that this response was "ordinary"? One step at a time.

Imagine that you are a therapist who cares deeply about the rights and safety of women and children. You see yourself as a skillful, compassionate practitioner. You know how hard it has been to get politicians and the public to pay serious attention to the problems of women and children. You know how difficult it has been for battered women to speak up. Now you start hearing about a new phenomenon: In therapy, women are suddenly recovering memories that they had repressed all their lives, memories of horrific events. These cases are turning up on talk shows, at the conferences you go to, and in a flurry of books, notably the hugely popular *The Courage to Heal* (1988). It's true that the book's authors, Ellen Bass and Laura Davis, have no training in any kind of psychological research or psychotherapy, let alone science, something they freely admitted. "None of what is presented here is based on psychological theories," Bass explained in the preface, but this ignorance of psychology did not prevent them from defining themselves as healers and experts on sexual abuse, based on the workshops they had led.[38] They provided a list of symptoms, any of which, they said, suggests that a woman may have been a victim of incest, including these: She feels powerless and unmotivated; she has an eating disorder or sexual problems; she feels there is something wrong with her deep down inside; she feels she has to be perfect; she feels bad, dirty, or ashamed. You are a therapist working with women who have some of these problems. Should you assume that years of incest, repressed from memory, is the primary cause?

There you are, at the top of the pyramid, with a decision to make: Leap onto the recovered-memory bandwagon or be skeptical. The majority of mental-health professionals chose the latter course and did not go along. But a large number of therapists—between one-fourth and one-third, according to several surveys[39]—took that first step in the direction of belief, and, given the closed loop of clinical practice, we can see how easy it was for them to do so. Most had not been trained in the show-me-the-data spirit of skepticism. They did not know about the confirmation bias, so it did not occur to them that Bass and Davis were seeing evidence of incest in any symptom a woman had and even in the fact that she had no symptoms. They lacked a deep appreciation of the importance of control groups, so they were unlikely to wonder how many women who were not molested nonetheless had eating disorders or felt powerless and unmotivated.[40] They did not pause to consider what reasons other than incest might cause their female clients to have sexual problems.

Even some skeptical practitioners were reluctant to slow the bandwagon by saying anything critical of their colleagues or the women telling their stories. It's uncomfortable — dissonant — to realize that some of your colleagues are tainting your profession with silly or dangerous ideas. It's embarrassing — dissonant — to realize that not everything women and children say is true, especially after all your efforts to persuade victimized women to speak up and get the world to recognize the problem of child abuse. Some therapists feared that to publicly question the veracity of recovered memories was to undermine the credibility of the women who really had been molested or raped. Some feared that criticism of the recovered-memory movement would give ammunition and moral support to sexual predators and antifeminists. In the beginning, they could not have anticipated that a national panic about sexual abuse would erupt and that innocent people would be swept up in the pursuit of the guilty. Yet by remaining silent as this happened, they furthered their own slide down the pyramid.

· · ·

What is the status today of recovered-memory therapy and its fundamental assumption that traumatic experiences are usually repressed? As sensational cases have faded from public attention, it might seem that the issues have been resolved, sanity and science having prevailed. But as dissonance theory predicts, once an incorrect idea has achieved prominence, and especially if that idea has caused widespread harm, it rarely fades away. It lies in wait, like the false belief that vaccines cause autism, reemerging at any opportunity that might allow its promoters to claim they were right all along. Lawsuits are still being filed, and families are still being broken apart by people who, in therapy, came to remember having been sexually molested or otherwise abused. The American Psychiatric Association changed the name *multiple personality disorder* to *dissociative identity disorder.* A professional association of trauma psychiatrists and psychotherapists who have promoted this diagnosis for years, under both of its labels, still gives its Cornelia Wilbur Award for "outstanding clinical contributions to the treatment of dissociative disorders."

A 2014 study reported that "although there are indications of more skepticism today than in the 1990s," the scientist-practitioner gap remains "a serious divide." The researchers sampled many groups of professional psychologists and psychotherapists and found that the more scientifically trained they were, the more accurate their beliefs about memory and trauma. Among members of the Society for a Science of Clinical Psychology, only 17.7 percent believed that "traumatic memories are often repressed." Among general psychotherapists, it was 60 percent; among psychoanalysts, 69 percent; and among neuro-linguistic programming therapists and hypnotherapists, 81 percent—which is about the same percentage found in the general public.[41] "The memory wars have not vanished," wrote a team of memory scientists in 2019. "They have continued to endure and contribute to potentially damaging consequences in clinical, legal, and academic contexts."[42]

No wonder this gap persists, given what it would take for all those psychotherapists and psychiatrists who generated the epidemic of recovered-memory and multiple-personality cases to climb back up the pyramid. Some continue to do what they have been doing for years, helping clients uncover "repressed" memories.[43] Others have quietly dropped their focus on repressed memories of incest as the leading explanation of their clients' problems; for them, it has gone out of fashion, just as penis envy, frigidity, and masturbatory insanity did decades ago. They drop one fad when it loses steam and sign on for the next, rarely pausing to question where all the repressed incest cases went. They might be vaguely aware that there is controversy, but it's easier to stay with what they have always done and maybe add a newer technique to go along with it. Some therapists have forgotten how enthusiastically they once believed in recovered-memory assumptions and methods and now see themselves as moderates in the whole debate.

Undoubtedly, the practitioners who have the greatest dissonance to resolve are the clinical psychologists and psychiatrists who most actively promoted and benefited from recovered-memory and multiple-personality therapies to begin with. Many have impressive credentials. The movement gave them great fame and success. They were star lecturers at professional conferences. They were (and still are) called on to testify in court about whether a child has been abused or whether a plaintiff's recovered memory is reliable, and, as we saw, they usually made their judgments with a high degree of confidence. As the scientific evidence that they were wrong began to accumulate, how likely was it that they would embrace the data readily, grateful for the studies of memory and children's testimony that would improve their practice? To do so would have been to realize that they had harmed the very women and children they said they were trying to help. It was much easier for them to preserve their commitments by rejecting the scientific research as being irrelevant to clinical practice. And as soon as they took that self-justifying step, they could not go back without enormous psychological difficulty.

Today, standing at the bottom of the pyramid, miles away professionally from their scientific colleagues and having devoted more than two decades to promoting a form of therapy that Richard McNally calls "the worst catastrophe to befall the mental-health field since the lobotomy era,"[44] most recovered-memory clinicians remain as committed as ever to their beliefs, continuing to preach what they have long practiced. How have they reduced their dissonance?

One popular method is by minimizing the extent of the problem and the damage it caused. Clinical psychologist John Briere, one of the earliest supporters of recovered-memory therapy, finally admitted at a conference that the large number of recovered-memory cases reported in the 1980s may have been caused, at least in part, by "overenthusiastic" therapists who had inappropriately tried to "liposuction memories out of their [clients'] brains." Mistakes were made — by them. But only a few of them, he hastened to add. Recovered false memories are rare, he said; repressed true memories are far more common.[45]

Others reduce dissonance by blaming the victim. Colin Ross, a psychiatrist who rose to fame and fortune by claiming that repressed memories of abuse cause multiple personality disorder, eventually agreed that "suggestible individuals can have memories elaborated within their minds because of poor therapeutic technique." But because "normal human memory is highly error-prone," he concluded that "false memories are biologically normal and, therefore, not necessarily the therapist's fault." Therapists don't create false memories in their clients, because therapists are merely "consultants."[46] Therefore, if a client comes up with a mistaken memory, it's the client's fault. (Colin Ross won the Cornelia Wilbur Award in 2016.)

The most ideologically committed clinicians reduced dissonance by killing the messenger. In the late 1990s, when psychiatrists and psychotherapists were being convicted of malpractice for their use of coercive methods to generate false recovered memories and multiple personalities, D. Corydon Hammond advised his clinical colleagues at a convention thus: "I think it's time somebody called for an open season on academicians and researchers. In the United States and Canada in particular, things have become so extreme with academics supporting extreme false memory positions, so I think it's time for clinicians to begin bringing ethics charges for scientific malpractice against researchers, and journal editors — most of whom, I would point out, don't have malpractice coverage."[47] Some psychiatrists and clinical psychologists took Hammond's advice and sent harassing letters to researchers and journal editors, made spurious claims of ethics violations against scientists studying memory and children's testimony, and filed nuisance lawsuits aimed at blocking publication of critical articles and books.[48] None of these efforts were successful at silencing the scientists.[49]

There is one final way they can reduce dissonance: Dismiss all of that scientific research as being part of a backlash against child victims and incest survivors. The concluding section of the third edition of *The Courage to Heal* is called "Honoring the Truth: A Response to the Backlash." There was no section called "Honoring the Truth: We Made Some Big Mistakes."[50]

• • •

There are almost no psychotherapists who practiced recovered-memory therapy, no child experts who sent the dozens of Bernard Barans to prison, who have admitted that they were wrong. From those few who have publicly admitted their errors, though, we can see what it took to shake them out of their protective cocoons of self-justification. For Linda Ross, it was taking herself out of the closed loop of private therapy sessions and forcing herself to confront, in person, parents whose lives had been destroyed by their grown children's accusations. One of her clients brought her to a meeting of accused parents. Ross suddenly realized that a story that had seemed bizarre but possible when her client told it in therapy now seemed fantastical when multiplied by a roomful of similar tales. "I had been so supportive of women and their repressed memories," she said, "but I had never once considered what that experience was like for the parents. Now I heard how absolutely ludicrous it sounded. One elderly couple introduced themselves, and the wife told me that their daughter had accused her husband of murdering three people . . . The pain in these parents' faces was so obvious. And the unique thread was that their daughters had gone to [recovered-memory] therapy. I didn't feel very proud of myself or my profession that day."

After that meeting, Ross said, she would frequently wake up in the middle of the night "in terror and anguish" as the cocoon began to crack open. She worried about being sued, but most of the time she "just thought about those mothers and fathers who wanted their children back." She called her former clients in an attempt to undo the damage she had caused, and she changed the way she practiced therapy. In an interview on National Public Radio's *This American Life* with Alix Spiegel, Ross told of accompanying one of her clients to a meeting with the woman's parents, whose home had been dismantled by police trying to find evidence of a dead body that their daughter had claimed to remember in therapy.[51] There was no dead body, any more than there were underground torture chambers at the McMartin Preschool. "So I had a chance to tell them the part that I played," said Ross. "And to tell them that I completely understood that they would find it difficult for the rest of their lives to be able to find a place to forgive me, but that I was certainly aware that I was in need of their forgiveness."

At the end of the interview, Alix Spiegel said: "There are almost no people like Linda Ross, practicing therapists who have come forward to talk publicly about their experience, to admit culpability, or try to figure out how this happened. The experts, for once, are strangely silent."

5

Law and Disorder

I guess it's really difficult for any prosecutor [to acknowledge errors and] to say, "Gee, we had 25 years of this guy's
life. That's enough."
— *Dale M. Rubin, lawyer for Thomas Lee Goldstein*

Thomas Lee Goldstein, a college student and ex-Marine, was
convicted in 1980 of a murder he did not commit, and he spent
the next twenty-four years in prison. His only crime was being
in the wrong place at the wrong time. Although he lived near
the murder victim, the police found no physical evidence linking
Goldstein to the crime — no gun, no fingerprints, no blood. He
had no motive. He was convicted on the testimony of a jailhouse
informant, improbably named Edward Fink, who had been arrested thirty-five times, had three felony convictions and a her-
oin habit, and had testified against ten different men, stating in
each case that the defendant had confessed to him while the two
were sharing a jail cell. (A prison counselor had described Fink
as "a con man who tends to handle the facts as if they were elas-

tic.") Fink lied under oath, denying that he had been given a reduced sentence in exchange for his testimony. The prosecution's
only other support for its case was an eyewitness, Loran Campbell, who identified Goldstein as the killer after the police falsely
assured him that Goldstein had failed a lie-detector test. None
of the other five eyewitnesses identified Goldstein, and four of
them said the killer was "black or Mexican." Campbell recanted
his testimony later, saying he had been "a little overanxious" to
help the police by telling them what they wanted to hear. It was
too late. Goldstein was sentenced to twenty-seven years to life
for the murder.

Over the years, five federal judges agreed that prosecutors
had denied Goldstein his right to a fair trial by failing to tell the
defense about their deal with Fink, but Goldstein remained in
prison. Finally, in February 2004, a California Superior Court
judge dismissed the case "in furtherance of justice," citing its lack
of evidence and its "cancerous nature" — its reliance on a professional informer who'd perjured himself. Even then, the Los Angeles County prosecutors refused to acknowledge that they might
have made a mistake. Within hours of the judge's decision, they
filed new charges against Goldstein, set bail at one million dollars, and announced they would retry him for the murder. "I am
very confident we have the right guy," deputy district attorney
Patrick Connolly said. Two months later, the DA's office con-
ceded it had no case against Goldstein and released him.

•   •   •

On the night of April 19, 1989, the woman who came to be known
as the Central Park Jogger was brutally raped and bludgeoned.

The police quickly arrested five black and Hispanic teenagers from Harlem who had been in the park "wilding," randomly attacking and roughing up passersby. The police, not unreasonably, saw them as likely suspects for the attack on the jogger. They kept the teenagers in custody and interrogated them intensively for fourteen to thirty hours. The boys, ages fourteen to sixteen, finally confessed to the crime, but they did more than admit guilt: They reported lurid details of what they had done. One boy demonstrated how he had pulled off the jogger's pants. One told how her shirt was cut off with a knife and how one of the gang repeatedly struck her head with a rock. Another expressed remorse for his "first rape," saying he had felt pressured by the other guys to do it and promising he would never do it again. There was no physical evidence linking the teenagers to the crime — no matching semen, blood, or DNA — and the prosecution *knew* that the DNA found on the victim did not match that of any of the teenagers'. But the boys' confessions persuaded the police, the jury, forensic experts, and the public that the perpetrators had been caught. Donald Trump spent eighty thousand dollars on newspaper ads calling for them to get the death penalty.[1]

And yet the teenagers were innocent. Thirteen years later, a felon named Matias Reyes, in prison for three rape-robberies and one rape-murder, admitted that he, and he alone, had committed the crime. He revealed details that no one else knew, and his DNA matched the DNA taken from semen found in the victim and on her sock. The Manhattan District Attorney's office, headed by Robert M. Morgenthau, investigated for nearly a year and could find no connection between Reyes and the boys who had been convicted. "If only we had DNA 13 years ago," he later lamented. His office supported the defense motion to vacate the boys' convictions, and in 2002 the motion was granted. It took another twelve years before New York City, without admitting error, reached a settlement with the Central Park Five for forty-one million dollars.

Morgenthau's decision was angrily denounced by former prosecutors in his office and by the police officers who had been involved in the original investigation; they refused to believe that the boys were innocent.[2] After all, they had confessed. The prosecutor in the case, Linda Fairstein, who was head of the sex crimes unit in the DA's office, had successfully prosecuted many heinous cases and was not disposed to consider that the Central Park Five might be innocent. She was so zealous in getting coerced confessions from the teenagers that an appellate court judge later singled her out in his opinion, noting, "I was concerned about a criminal justice system that would tolerate the conduct of the prosecutor, Linda Fairstein, who deliberately engineered the 15-year-old's confession." In 2004, two years after Matias Reyes was indisputably identified as the rapist and after the five young men were released from prison, Fairstein told a reporter that she was certain that the original convictions were correct: "Those of us on the prosecution team have always been looking for the sixth man," she said. "I think [the five] were freed because it was politically expedient to do so."[3] Neither Sarah and Ken Burns's blistering 2012 documentary *The Central Park Five* nor Ava DuVernay's 2019 dramatized account *When They See Us* changed her mind. "Ava DuVernay's miniseries wrongly por-

trays them as totally innocent — and defames me in the process,"
Fairstein wrote in an op-ed for the *Wall Street Journal.*[4]

Fairstein retired to write novels featuring an intrepid DA ("a
younger, thinner, blonder me," she said) who always gets her man
— a creative way of reducing dissonance indeed.

• • •

In 1932, Yale law professor Edwin Borchard published *Convicting
the Innocent: Sixty-Five Actual Errors of Criminal Justice.* Eight
of those cases involved defendants who'd been convicted of murder even though the supposed victim had turned up later, very
much alive. You'd think that might be fairly convincing proof
that police and prosecutors had made some serious mistakes, yet
one prosecutor told Borchard, "Innocent men are never convicted. Don't worry about it, it never happens . . . It is a physical
impossibility."

Then came DNA. Ever since 1989, the first year DNA testing resulted in the release of an innocent prisoner, the public has
been repeatedly confronted with evidence that, far from being
an impossibility, convicting the innocent is much more common than we feared. The Innocence Project, founded by Barry
Scheck and Peter J. Neufeld, keeps a running record on its website of the hundreds of people imprisoned for murder or rape
who have been cleared by DNA testing; by 2019, the number
was 365.[5]

Understandably, wrongful convictions that are overturned
by DNA evidence get a great deal of public attention. But, as
we will see, DNA evidence isn't always relevant; people can be

wrongfully convicted for many reasons, from prosecutorial zeal and misconduct to junk-science testimony to faulty eyewitness testimony. Estimates of the rate of false convictions in the United States range from half of 1 percent at the low end to 2 to 3 percent at the high end. Law professor Samuel R. Gross, a national expert on exonerations, put his own estimate of wrongful convictions for felonies somewhat higher, from 1 to 5 percent. "Is that a lot or a little?" he wrote.

> That depends on your point of view. If as few as 1% of serious felony convictions are erroneous, that means that perhaps ten- to twenty-thousand or more of the nearly 2.3 million inmates in American prisons and jails are innocent. If as few as 1/10 of 1% of jetliners crashed on takeoff, we would shut down every airline in the country. That is not a risk we are prepared to take, and we believe we know how to address that sort of problem. Are 10,000 to perhaps 50,000 wrongfully imprisoned citizens too many? Can we do better? How? There are no obvious answers. The good news is that the great majority of convicted criminal defendants in America are guilty. The bad news is that a substantial number are not.[6]

In 2012, Gross and Rob Warden, the executive director of the Center on Wrongful Convictions at the Northwestern University School of Law, launched the National Registry of Exonerations, DNA- and non-DNA-determined. As its website notes, "The Registry provides detailed information about every known exoneration in the United States since 1989 — cases in which a person was wrongly convicted of a crime and later cleared of all the charges based on new evidence of innocence. The Registry also maintains a more limited database of known exonerations prior to 1989." Within two years, they had recorded more than fourteen hundred exonerations — more than four times the number of innocent people freed by DNA testing — and as of 2019, the number was nearly twenty-five hundred. The registry counts as exonerations those cases in which a person who has been convicted of a crime is officially cleared based on new evidence of innocence. It excludes the many cases in which a person convicted of a crime has been cleared for reasons that did not involve "new evidence of innocence." Therefore, as the website states, "The exonerations we know about are just a fraction of those that have taken place."[7]

This is uncomfortably dissonant information for anyone who wants to believe that the system works. Resolving it is hard enough for the average citizen, but if you are a participant in the justice system, your motivation to justify its mistakes, let alone yours, will be immense. Social psychologist Richard Ofshe, an expert on the psychology of false confessions, observed that convicting the wrong person is "one of the worst professional errors you can make — like a physician amputating the wrong arm."[8]

Suppose that you are presented with evidence that you did the legal equivalent of amputating the wrong arm: you helped send the wrong person to prison. What do you do? Your first impulse will be to deny your mistake for the obvious reason of protecting your job, reputation, and colleagues. Besides, if you release someone who later commits a serious crime or even if you free someone who is innocent but who was erroneously imprisoned for a heinous crime such as child molesting, an outraged public may nail you for it.[9] You have plenty of external incentives for denying that you made a mistake, but you have an even greater internal one: You want to think of yourself as an honorable, competent person who would never help convict the wrong guy. But how can you possibly think you got the right guy in the face of the new evidence to the contrary? Because, you assure yourself, the evidence is lousy, and look, he's a bad guy; even if he didn't commit this particular crime, he undoubtedly committed another one. The alternative, that you sent an innocent man to prison for fifteen years, is so antithetical to your view of your competence that you will jump through multiple mental hoops to convince yourself that you couldn't possibly have made such a blunder.

With each innocent person freed from years in prison through DNA testing, the public can almost hear the mental machinations of prosecutors, police, and judges who are busy resolving dissonance. One strategy is to claim that most of those cases don't reflect wrongful *convictions* but wrongful *pardons:* just because a prisoner is exonerated doesn't invariably mean he or she is innocent. And if the person really is innocent, well, that's a shame, but wrongful convictions are extremely rare, a reasonable price to pay for the superb system we have in place. The real problem is that too many criminals get off on technicalities or escape justice because they are rich enough to buy high-priced defense teams. As Joshua Marquis, a former Oregon district attorney and something of a professional defender of the criminal justice system, put it, "Americans should be far more worried about the wrongfully freed than the wrongfully convicted."[10]

When the nonpartisan Center for Public Integrity published its report of 2,012 cases of documented prosecutorial misconduct that led to wrongful convictions, Marquis dismissed the numbers and the report's implication that the problem might be "epidemic." "The truth is that such misconduct is better described as episodic," he wrote, "those few cases being rare enough to merit considerable attention by both the courts and the media."

Sadly, they are hardly rare. According to the Center for Prosecutor Integrity, established in 2014, there have been an estimated 16,000 cases of prosecutorial misconduct since 1970; fewer than 2 percent have resulted in any punishment for the offending prosecutor. A comprehensive analysis of 707 cases of confirmed misconduct in California from 1997 through 2009 showed that the courts set aside the conviction or sentence or declared a mistrial in only about 20 percent of them. And only 1 percent of the prosecutors who committed misconduct were publicly disciplined by the state bar. The report concluded that "prosecutors continue to engage in misconduct, sometimes multiple times, almost always without consequence. And the courts' reluctance to report prosecutorial misconduct and the State Bar's failure to discipline it" allows prosecutors to get away with murder — in the case of falsely convicting innocent men, sometimes literally.[11]

This evidence does not deter defenders of the status quo. When mistakes or misconduct occur, they maintain, the system has many self-correcting procedures in place to fix them immediately. In fact, Marquis worries that if we start tinkering with the system to reduce the rate of wrongful convictions, we will end up freeing too many guilty people. This claim reflects the perverted logic of self-justification. When an innocent person is falsely convicted, the *real* guilty party remains on the streets. "Alone among the legal profession," Marquis claimed, "a prosecutor's sole allegiance is to the truth — even if that means torpedoing the prosecutor's own case."[12] That is an admirable, dissonance-reducing sentiment, one that reveals the underlying problem more than Marquis realizes. It is precisely because prosecutors believe they are pursuing the truth that they do not torpedo their own cases when they need to; thanks to self-justification, they rarely think they need to.

You do not have to be a scurrilous, corrupt DA to think this way. Rob Warden has observed dissonance at work among prosecutors whom he considers "fundamentally good" and honorable people who want to do the right thing. When one exoneration took place, Jack O'Malley, the prosecutor on the case, kept saying to Warden, "How could this be? How could this happen?" Warden told a reporter, "He didn't get it. He didn't understand. He really didn't. And Jack O'Malley was a good man." Yet prosecutors cannot get beyond seeing themselves and the cops as good guys and the defendants as bad guys. "You get in the system," Warden said, "and you become very cynical. People are lying to you all over the place. Then you develop a theory of the crime, and it leads to what we call tunnel vision. Years later overwhelming evidence comes out that the guy was innocent. And you're sitting there thinking, 'Wait a minute. Either this overwhelming evidence is wrong or I was wrong — and I couldn't have been wrong because I'm a good guy.' That's a psychological phenomenon I have seen over and over."[13]

That phenomenon is self-justification. Over and over, as the two of us read the research on wrongful convictions in American history, we saw how self-justification can escalate the likelihood of injustice at every step of the process, from capture to conviction. The police and prosecutors use methods gleaned from a lifetime of experience to identify a suspect and build a case for conviction. Usually, they are right. Unfortunately, those same methods increase their risks of pursuing the wrong suspect, ignoring evidence that might implicate another, reinforcing their commitment to a wrong decision, and, later, refusing to admit their error. As the process rolls along, those who are caught up in the effort to convict the original suspect often become more certain that they have the perpetrator and more committed to getting a conviction. Once that person goes to jail, they think, that fact alone justifies what they did to put him there. Besides, the judge and jury agreed, didn't they? Self-justification not only puts innocent people in prison, therefore, but sees to it that they stay there.

The Investigators

On the morning of January 21, 1998, in Escondido, California, twelve-year-old Stephanie Crowe was found in her bedroom, stabbed to death. The night before, neighbors had called 911 to report their fears about a vagrant in the neighborhood who was behaving strangely — a man named Richard Tuite, who suffered from schizophrenia and had a history of stalking young women and breaking into their houses. But Escondido detectives and a team from the FBI's Behavioral Analysis Unit concluded almost immediately that the killing was an inside job. They knew that most murder victims are killed by someone related to them, not by crazy intruders.

Accordingly, the detectives, primarily Ralph Claytor and Chris McDonough, turned their attention to Stephanie's brother, Michael, then age fourteen. Michael, who was sick with a fever, was interrogated, without his parents' knowledge, for three hours at one sitting and then for another six hours without a break. The detectives lied to him; they said they'd found Stephanie's blood in his room, that she'd had strands of his hair in her hand, that someone inside the house had to have killed her because all the doors and windows were locked, that Stephanie's blood was all over his clothes, and that he had failed the computerized voice stress analyzer. (This is a pseudoscientific technique that supposedly identifies liars by measuring microtremors in their voices. No one has scientifically demonstrated the validity of this method.[14]) Although Michael repeatedly told them he had no memory of the crime and provided no details, such as where he put the murder weapon, he finally confessed that he had killed her in a jealous rage. Within days, the police also arrested Michael's friends Joshua Treadway and Aaron Houser, both fifteen. Joshua Treadway, after two interrogations that lasted twenty-two hours, produced an elaborate story of how the three of them had conspired to murder Stephanie.

On the eve of the trial, in a dramatic turn of events, Stephanie's blood was discovered on the sweatshirt that Richard Tuite, the vagrant, had been wearing the night of her murder. This evidence forced district attorney Paul Pfingst to dismiss the charges against the teenagers, although, he said, he remained convinced

of their guilt because of their confessions and he would therefore not indict Tuite. The detectives who had pursued the boys,
Claytor and McDonough, never gave up their certainty that
they had nabbed the real killers. They self-published a book to
justify their procedures and beliefs. In it, they claimed that Richard Tuite was just a fall guy, a scapegoat, a drifter who had been
used as a pawn by politicians, the press, celebrities, and the criminal and civil lawyers hired by the boys' families to "shift blame
from their clients and transfer it to him instead."[15]

The teenagers were released and the case was handed over
to another detective in the department, Vic Caloca, to dispose
of. Despite opposition by the police and the district attorneys,
Caloca reopened the investigation. Other cops stopped talking
to him; a judge scolded him for making waves; the prosecutors
ignored his requests for assistance. He had to get a court order to
get evidence he sought from a crime lab. Caloca persisted, eventually compiling a three-hundred-page report listing the "spec-
ulations, misjudgments and inconclusive evidence" used in the
case against Michael Crowe and his friends. Because Caloca
was not part of the original investigating team and so had not
jumped to the wrong conclusion, the evidence implicating Tuite
was not dissonant for him. It was simply evidence.

Caloca bypassed the local DA's office and took that evidence
to the California State Attorney General's office in Sacramento.
There, assistant attorney general David Druliner agreed to prosecute Tuite. In May 2004, six years after the investigating detectives had ruled him out as a suspect, deciding he was nothing
more than a bungling prowler, Richard Tuite was convicted of

the murder of Stephanie Crowe.* Druliner was highly critical of the initial investigation by the Escondido detectives. "They went off completely in the wrong direction to everyone's detriment," he said. "The lack of focus on Mr. Tuite — we could not understand that."[16]

Yet by now the rest of us can. It does seem ludicrous that the detectives did not change their minds, or at least entertain doubts for a moment, when Stephanie's blood turned up on Tuite's sweater. But once the detectives had convinced themselves that Michael and his friends were guilty, they started down the decision pyramid, self-justifying every bump to the bottom.

Let's begin at the top, with the initial process of identifying a suspect. Many detectives do just what the rest of us are inclined to do when we first hear about a crime — we impulsively decide we know what happened and then fit the evidence to support our conclusions, ignoring or discounting evidence that contradicts it. Social psychologists have studied this phenomenon extensively by putting people in the role of jurors and seeing what factors influence their decisions. In one experiment, mock jurors listened to an audiotaped reenactment of an actual murder trial and then said how they would have voted and why. Instead of considering and weighing possible verdicts in light of the evidence, most people immediately constructed a story about what had happened and then, as evidence was presented during the re-

---

* In 2013, Tuite's conviction was overturned on a technicality, and at his subsequent trial, the jury ruled that the lackluster prosecution had failed to prove his guilt beyond a reasonable doubt.

enactment of the trial, they accepted only the evidence that supported their preconceived version of what had happened. Those who jumped to a conclusion early on were the most confident in their decisions and the most likely to justify it by voting for an extreme verdict.[17] This is normal; it's also alarming.

In their first interview with a suspect, detectives tend to make a snap decision: Is this guy guilty or innocent? Over time and with experience, the police learn to pursue certain leads and reject others, eventually becoming certain of their accuracy. Their confidence is partly a result of experience and partly a result of training techniques that reward speed and certainty over caution and doubt. Jack Kirsch, a former chief of the FBI's Behavioral Science Unit, told an interviewer that visiting police officers often came up to his team members with difficult cases and asked for advice. "As impromptu as it was, we weren't afraid to shoot from the hip and we usually hit our targets," he said. "We did this thousands of times."[18]

This confidence is often well placed, because usually the police are dealing with confirming cases, the people who are guilty. Yet it also raises the risks of mislabeling the innocent as guilty and thereby shutting the door on other possible suspects too soon. Once that door closes, so does the mind. Thus, the detectives didn't even try using their fancy voice analyzer on Tuite as they had on Crowe. Detective McDonough explained that "since Tuite had a history of mental illness and drug use, and might still be both mentally ill and using drugs currently, the voice stress testing might not be valid."[19] In other words, "Let's use our unreliable gizmo only on suspects we already believe are guilty, because whatever they do, it will confirm our belief; we won't use it on suspects we believe are innocent, because it won't work on them anyway."

The initial decision about a suspect's guilt or innocence appears obvious and rational at first: The suspect may fit a description given by the victim or an eyewitness, or the suspect may fit into a statistically likely category. Follow the trail of love and money, and the force is with you. Thus, in the majority of murders, the most probable killer is the victim's lover, spouse, ex-spouse, relative, or beneficiary. Lieutenant Ralph M. Lacer was therefore certain that a Chinese American college student named Bibi Lee had been killed by her boyfriend, Bradley Page, which was why he did not follow up on testimony from eyewitnesses who had seen a man near the crime scene push a young "Oriental" woman into a van and drive away.[20] When a young woman is murdered, said Lacer, "The number one person you're going to look for is her significant other. You're not going to be looking for some dude out in a van." However, as attorney Steven Drizin observes, "Family members may be a legitimate starting point for an investigation but that's all they are. Instead of trying to prove the murder was intra-family, police need to explore all possible alternatives. All too often they do not."[21]

Once a detective decides that he or she has found the killer, the confirmation bias sees to it that the prime suspect becomes the only suspect. And if the prime suspect happens to be innocent, too bad — he's still on the ropes. In the introduction, we described the case of Patrick Dunn, who was arrested in Kern County, California, and charged with murdering his wife. In that case, the police chose to believe a career criminal's uncorroborated account of events, which supported their theory that

Dunn was guilty, rather than the corroborated statements by an impartial witness, which weakened it. This decision was unbelievable to the defendant, who asked his lawyer, Stan Simrin, "But don't they want the truth?" "Yes," Simrin said, "and they are convinced they have found it. They believe the truth is you are guilty. And now they will do whatever it takes to convict you."[22]

Doing whatever it takes to convict someone leads to ignoring or discounting evidence that would require officers to change their minds about a suspect. In extreme cases, it can tempt individual officers and even entire departments to cross the line from legal to illegal actions. The Rampart Division of the Los Angeles Police Department set up an antigang unit in which dozens of officers were eventually charged with making false arrests, giving perjured testimony, and framing innocent people; nearly one hundred convictions that had been attained using these illegal methods were overturned. And in New York, a state investigation found that the Suffolk County Police Department had botched a number of major cases by brutalizing suspects, illegally tapping phones, and losing or faking crucial evidence.

Corrupt officers like these are made, not born. They are led down the slope of the pyramid by the culture of the police department and by their own loyalty to its goals. Law professor Andrew McClurg has traced the process that leads many officers to behave in ways they never would have imagined when they started out as idealistic rookies. Being called on to lie in the course of their official duties at first creates dissonance, "I'm here to uphold the law" versus "And here I am, I'm breaking it myself." Over time, observes McClurg, they "learn to smother their dissonance under a protective mattress of self-justification."

Once officers believe that lying is defensible and even an essential aspect of the job, he adds, "dissonant feelings of hypocrisy no longer arise. The officer learns to rationalize lying as a moral act or at least as not an immoral act. Thus, his self-concept as a decent, moral person is not substantially compromised."[23]

Let's say you're a cop serving a search warrant on a rock house, a place where crack cocaine is sold. You chase one guy to the bathroom, hoping to catch him before he flushes the dope, and your case, down the drain. You're too late. There you are, revved up, adrenaline flowing, you've put yourself in harm's way — and this bastard is going to get away? Here you are in a rock house, you and your partner know what is going on, and these scumbags are going to walk? They are going to get a slick lawyer, and they will be out in a heartbeat. All that work, all that risk, all that danger — for nothing? Why not take a little cocaine out of your pocket, the stuff from that bust earlier today, drop it on the floor of that bathroom, and nail the perp with it? All you'd have to say is "Some of that crack fell out of his pocket before he could flush it all."[24]

It's easy to understand why you would do this, under the circumstances. It's because you want to do your job. You know it's illegal to plant evidence, but it seems so justifiable. The first time you do it, you tell yourself, *The guy is guilty!* This experience will make it easier for you to do the same thing again; in fact, you will be strongly motivated to repeat the behavior, because to do otherwise is to admit, if only to yourself, that it was wrong the first time you did it. Before long, you are breaking the rules in more ambiguous situations. Because police culture generally supports these justifications, it becomes even harder for an individual officer to resist breaking (or bending) the rules. Eventually, many cops will take the next steps, proselytizing other officers, persuading them to join them in a little innocent rule-breaking, and shunning or sabotaging officers who do not go along—and who are a reminder of the moral road not taken.

And, in fact, the 1992 Mollen Commission, reporting on patterns of corruption in the New York Police Department, concluded that the practice of police falsification of evidence is "so common in certain precincts that it has spawned its own word: 'testilying.'"[25] In such police cultures, police routinely lie to justify searching anyone they suspect of having drugs or guns, swearing in court that they stopped a suspect because his car ran a red light, because they saw drugs changing hands, or because the suspect dropped the drugs as the officer approached, giving him probable cause to arrest and search the guy.* Norm Stamper, a police officer for thirty-four years and former chief of the Seattle Police Department, has written that there isn't a major police force in the country "that has escaped the problem: cops, sworn to uphold the law, [are] seizing and converting drugs to their own use [and] planting dope on suspects."[26] The most common justification for lying and planting evidence is that the end justifies the means. One officer told the Mollen Commission investigators that he was "doing God's work." Another said, "If we're

---

* The practice doesn't change, only the rules of what police are permitted or forbidden to do. In July 2019, a New York judge issued a scathing opinion condemning the "ubiquitous" police practice of claiming that they "smelled an odor of marijuana" to justify an otherwise illegal stop-and-search. "The time has come to reject the canard of marijuana emanating from nearly every vehicle subject to a traffic stop," she wrote. See Joseph Goldstein, "Officers Said They Smelled Pot. A Judge Called Them Liars," *New York Times,* September 13, 2019.

going to catch these guys, fuck the Constitution." When one officer was arrested on charges of perjury, he asked in disbelief, "What's wrong with that? They're guilty."[27]

What's wrong with that is that there is nothing to prevent the police from planting evidence and committing perjury to convict someone they believe is guilty — but who is innocent. Corrupt cops are certainly a danger to the public, but so are many of the well-intentioned ones who would never dream of railroading an innocent person into prison. In a sense, honest cops are even more dangerous than corrupt cops, because they are far more numerous and harder to detect. The problem is that once they have decided on a likely suspect, they don't think it's possible that he or she is innocent. Therefore, once they have a suspect, they behave in ways to confirm that initial judgment of guilt, justifying the techniques they use in the belief that only guilty people will be vulnerable to them.

The Interrogators

The most powerful piece of evidence a detective can produce in an investigation is a confession, because it is the one thing most likely to convince a prosecutor, jury, and judge of a person's guilt. Accordingly, police interrogators are trained to get that confession, even if that means lying to the suspect and using, as one detective proudly admitted to a reporter, "trickery and deceit."[28] Most people are surprised to learn that this is entirely legal. Detectives are proud of their ability to trick a suspect into confessing; it's a mark of how well they have learned their trade. The greater their confidence, the greater the dissonance they will feel if confronted with evidence that they were wrong, and the greater the need to reject that evidence.

Inducing an innocent person to confess is obviously the most dangerous mistake that can occur in police interrogation, but most detectives, prosecutors, and judges don't think it is possible. "The idea that somebody can be induced to falsely confess is ludicrous," said Joshua Marquis. "It's the Twinkie defense of [our time]. It's junk science at its worst."[29] Most of us agree, because we can't imagine we would admit to committing a crime if we were innocent. We'd protest. We'd stand firm. We'd call for our lawyer ... wouldn't we? Yet the National Registry's list of unequivocally exonerated prisoners notes that about 13 to 15 percent of them had confessed to crimes they had not committed. Social scientists and criminologists have analyzed many of these cases and conducted experimental research to demonstrate how this can happen.

The bible of interrogation methods is *Criminal Interrogation and Confessions,* written by Fred E. Inbau, John E. Reid, Joseph P. Buckley, and Brian C. Jayne. John E. Reid and Associates offers training programs, seminars, and videos on the nine-step Reid Technique, and on their website, they claim that they have trained more than five hundred thousand law enforcement workers in the most effective ways of eliciting confessions. *Criminal Interrogation and Confessions* starts right off by reassuring readers that "we are opposed to the use of any interrogation tactic or technique that is apt to make an innocent person confess" even though some interrogations "require the use of psychological tactics and techniques that could well be classified as 'unethical,' if evaluated in terms of ordinary, everyday social behavior."[30]

> It is our clear position that merely introducing fictitious evidence during an interrogation would not cause an innocent person to confess. It is absurd to believe that a suspect who knows he did not commit a crime would place greater weight and credibility on alleged evidence than his own knowledge of his innocence. Under this circumstance, the *natural human reaction* would be one of anger and mistrust toward the investigator. The net effect would be the suspect's further resolution to maintain his innocence.[31]

Wrong. The "natural human reaction" is usually not anger and mistrust but confusion and hopelessness — dissonance — because most innocent suspects trust the investigator not to lie to them. The interrogator, however, is biased from the start. Whereas an interview is a conversation designed to get general information from a person, an interrogation is designed to get a suspect to admit guilt. (The suspect is often unaware of the difference.) The manual states this explicitly: "An interrogation is conducted only when the investigator is reasonably certain of the suspect's guilt."[32] The danger of that attitude is that once the investigator is "reasonably certain," the suspect cannot dislodge that certainty. On the contrary, anything the suspect does will be interpreted as evidence of lying, denial, or evading the truth, including repeated claims of innocence. Interrogators are explicitly instructed to think this way. They are taught to adopt the attitude "Don't lie; we know you are guilty" and to reject the suspect's denials. We've seen this self-justifying loop before, in the way some therapists and social workers interview children they believe have been molested. Once an interrogation like this has begun, there is no such thing as disconfirming evidence.[33]

Promulgators of the Reid Technique have an intuitive understanding of how dissonance works (at least in other people). They realize that if a suspect is given the chance to protest his innocence, he will have made a public commitment and it will be harder for him to back down later and admit guilt. "The more the suspect denies his involvement," writes Louis Senese, vice president of Reid and Associates, "the more difficult it becomes for him to admit that he committed the crime." Precisely — because of dissonance. Therefore, Senese advises interrogators to be prepared for the suspect's denials and head them off at the pass. Interrogators, he says, should watch for nonverbal signs that the suspect is about to deny culpability ("holding his hand up or shaking his head no or making eye contact"), and if the suspect says straight out, "Could I say something?," interrogators should respond with a command, using the suspect's first name ("Jim, hold on for just a minute") and then return to their questioning.[34]

The interrogator's presumption of guilt creates a self-fulfilling prophecy. It makes the interrogator more aggressive, which in turn makes innocent suspects behave more suspiciously. In one experiment, social psychologist Saul Kassin and his colleagues paired individuals who were either guilty or innocent of a mock theft with interrogators who were told that the suspects were guilty or that they were innocent. There were therefore four possible combinations of suspect and interrogator: You're innocent and he thinks you're innocent; you're innocent and he thinks you're guilty; you're guilty and he thinks you're innocent; and you're guilty and he thinks you're guilty. The deadliest combination, the one that produced the greatest degree of pressure and coercion by the interviewer, was the one that paired an interrogator convinced of a suspect's guilt with a suspect who was actually innocent. In such circumstances, the more the suspect denied guilt, the more certain the interrogator became that the suspect was lying, and he upped the pressure accordingly.

Kassin lectures widely to detectives and police officers to show them how their techniques of interrogation can backfire. They always nod knowingly, he says, and agree with him that false confessions are to be avoided. but then they immediately add that they themselves have never coerced anyone into a false confession. "How do you know?" Kassin asked one cop. "Because I never interrogate innocent people," he said. Kassin found that this certainty of infallibility starts at the top. "I was at an International Police Interviewing conference in Quebec, on a debate panel with Joe Buckley, president of the Reid School," he told us. "After his presentation, someone from the audience asked whether he was concerned that innocent people might confess in response to his techniques. Son of a gun if he didn't say it, word for word; I was so surprised at his overt display of such arrogance that I wrote down the quote and the date on which he said it: 'No, because we don't interrogate innocent people.'"[35] (In this he echoes the remarks of Ronald Reagan's attorney general Edwin Meese, who said in 1986, "But the thing is, you don't have many suspects who are innocent of a crime. That's contradictory. If a person is innocent of a crime, then he is not a suspect.")

In the next phase of training, detectives become confident in their ability to read the suspect's nonverbal cues: eye contact, body language, posture, hand gestures, and vehemence of denials. If the person won't look you in the eye, the manual explains, that's a sign of lying. If the person slouches (or sits rigidly), those are signs of lying. If the person denies guilt, that's a sign of lying. Yet the Reid Technique advises interrogators to "deny suspect eye contact." Deny a suspect the direct eye contact that they themselves regard as evidence of innocence?

The Reid Technique is thus a closed loop: How do I know a suspect is guilty? Because he's nervous and sweating (or too controlled) and because he won't look me in the eye (and I wouldn't let him if he wanted to). So my partners and I interrogate him for twelve hours using the Reid Technique, and he confesses. Therefore, because innocent people never confess, his confession confirms my belief that his being nervous and sweating (or too controlled) and looking me in the eye (or not) is a sign of guilt. By the logic of this system, the only error the detective can make is failing to get a confession.

The manual is written in an authoritative tone as if it were the voice of God revealing indisputable truths, but in fact it fails to teach its readers a core principle of scientific thinking: the importance of examining and ruling out other possible explanations for a person's behavior before deciding which one is the most likely. Saul Kassin was involved in a military case in which investigators had relentlessly interrogated a defendant against whom there was no hard evidence. (Kassin believed the man to be innocent, and indeed he was acquitted.) When one of the investigators was asked why he pursued the defendant so aggressively, he said: "We gathered that he was not telling us the whole truth. Some examples of body language is that he tried to remain calm, but you could tell that he was nervous and every time we tried to ask him a question his eyes would roam and he would not make direct contact, and at times he would act pretty sporadic and he started to cry at one time."

"What he described," says Kassin, "is a person under stress." Students of the Reid Technique generally do not learn that being nervous, fidgeting, avoiding eye contact, and slouching uncomfortably might be signs of something other than guilt. They might be signs of nervousness, adolescence, cultural norms, deference to authority — or anxiety about being falsely accused.

Promoters of the manual claim that their method trains investigators to determine whether someone is telling the truth or lying with an 80 to 85 percent level of accuracy. There is simply no scientific support for this claim. As with the psychotherapists we discussed in chapter 4, training does not increase accuracy; it increases people's confidence in their accuracy. In one of numerous studies that have documented the false-confidence phenomenon, Kassin and his colleague Christina Fong trained a group of students in the Reid Technique. The students watched the Reid training videos, read the manual, and were tested on what they had learned to make sure they got it. Then they were asked to watch videos of people being interviewed by an experienced police officer. The filmed suspects were either guilty of a crime but denying it, or denying it because they were innocent. The training did not improve the students' accuracy by an iota. They did no better than chance, but it did make them feel more confident of their abilities. Still, they were only college students, not professionals. So Kassin and Fong asked forty-four professional detectives in Florida and Ontario, Canada, to watch the tapes. These professionals averaged nearly fourteen years of experience each, and two-thirds had had special training, many in the Reid Technique. Like the students, they did no better than chance, yet they were convinced that their accuracy rate was close to 100 percent. Their experience and training did not improve their performance. Their experience and training simply increased their belief that it did.[36]

Nonetheless, why doesn't an innocent suspect just keep denying guilt? Why doesn't the target get angry at the interrogator, as the manual says any innocent person would do? Let's say you are an innocent person who is called in for questioning, perhaps to "help the police in their investigation." You have no idea that you are a prime suspect. You trust the police and want to be helpful. Yet here is a detective telling you that your fingerprints are on the murder weapon. That you failed a lie-detector test. That your blood was found on the victim or the victim's blood was on your clothes. These claims will create considerable cognitive dissonance:

COGNITION 1: I was not there. I didn't commit the crime. I have no memory of it.

COGNITION 2: Reliable and trustworthy people in authority tell me that my fingerprints are on the murder weapon, the victim's blood was on my shirt, and an eyewitness saw me in a place where I am sure I've never been.

How will you resolve this dissonance? If you are strong enough, wealthy enough, or have had enough experience with the police to know that you are being set up, you will say the four magic words: "I want a lawyer." But many people believe they don't need a lawyer if they are innocent.[37] Believing as they do (wrongly) that the police are not allowed to lie to them, they are astonished to hear that there is evidence against them. And what damning evidence at that — their fingerprints! The manual claims that the "self-preservation instincts of an innocent person during an interrogation" will override anything an interrogator does, but for vulnerable people, the need to make sense of what is happening to them trumps even the need for self-preservation.

BRADLEY PAGE: Is it possible that I could have done this terrible thing and blanked it out?

LIEUTENANT LACER: Oh, yes. It happens all the time.

And now the police offer you an explanation that makes sense, a way to resolve your dissonance: You don't remember because you blanked it out; you were drunk and lost consciousness; you repressed the memory; you didn't know that you had multiple personality disorder, and one of your other personalities did it. This is what the detectives did in their interrogations of Michael Crowe. They told him that there might have been "two Michaels," a good one and a bad one, and the bad Michael committed the crime without the good Michael even being aware of it.

But Michael was fourteen, you might say; no wonder the police could scare him into confessing. It is true that juveniles and the mentally ill are particularly vulnerable to these tactics, but so are healthy adults. In a close examination of 125 cases in which prisoners were later exonerated despite having given false confessions, Steven Drizin and Richard Leo found that more than half were not mentally ill, mentally deficient, or juveniles. Of the cases in which length of interrogation could be determined, more than 80 percent of the false confessors had been grilled for more than six hours straight, half for more than twelve hours, and some almost nonstop for two days.[38]

That was what happened to the teenagers arrested on the night the Central Park Jogger was attacked. They were interrogated for many hours without electronic recording of any kind, but the prosecutors subsequently created videos of brief recaps of the confessions for four of the five. When social scientists and legal scholars were able to examine all of the existing evidence, and when district attorney Robert Morgenthau's office reexamined this evidence starting from the assumption that the boys might be innocent rather than guilty, the dramatic persuasiveness of their confessions melted in the light. Their statements turned out to be full of contradictions, factual errors, guesses, and information planted by the interrogator's biased questions.[39] And contrary to the public impression that all of them confessed, in fact none of the defendants ever admitted that he personally raped the jogger. One said he "grabbed at" her. Another stated that he "felt her tits." One said he "held and fondled her leg." The district attorney's motion to vacate their convictions observed that "the accounts given by the five defendants differed from one another on the specific details of virtually every major aspect of the crime — who initiated the attack, who knocked the victim down, who undressed her, who struck her, who held her, who raped her, what weapons were used in the course of the assault, and when in the sequence of events the attack took place."[40]

After long hours of interrogation, wanting nothing more than to be allowed to go home, the exhausted suspect accepts the explanation the interrogators offer as the only one possible, the only one that makes sense. And confesses. Usually, once the pressure is over and the target gets a night's sleep, he or she will immediately retract the confession. It will be too late.

The Prosecutors

In that splendid film *The Bridge on the River Kwai,* Alec Guinness and his soldiers, prisoners of the Japanese in World War II, construct a railway bridge that will aid the enemy's war effort. Guinness agrees to this demand by his captors as a way of creating unity and restoring morale among his men, but once he builds it, it becomes *his* — a source of pride and satisfaction. When, at the end of the film, Guinness finds the wires revealing that the bridge has been mined and realizes that Allied commandos are planning to blow it up, his first reaction is, in effect: "You can't! It's my bridge. How dare you destroy it!" To the horror of the watching commandos, he tries to cut the wires to protect the bridge. Only at the very last moment does Guinness cry, "What have I done?," realizing that he was about to sabotage his own side's goal of victory to preserve his magnificent creation.

In the same way, many prosecutors end up prepared to sabotage their own side's goal of justice to preserve their convictions, in both meanings of the word. By the time prosecutors go to trial, they often find themselves in the real-world equivalent of a justification-of-effort experiment. They have selected this case out of many because they are convinced the suspect is guilty and that they have the evidence to convict. They have often invested many months preparing for it. They have worked intensely with police, witnesses, and the victim's shattered, often vengeful family. If the crime has roused public emotions, they are under enormous pressure to get a conviction quickly. Any doubts they might have are drowned in the satisfaction of feeling that they are representing the forces of good against a vile criminal. And so, with a clear conscience, prosecutors end up saying to a jury: "This defendant is subhuman, a monster. Do the right thing. Convict." Occasionally they so thoroughly convince themselves that they have caught a monster that they, like the police, go too far: coaching witnesses, offering deals to jailhouse informants, or failing to give the defense all the information they are legally obliged to hand over.

How, then, will most prosecutors react when, years later, the convicted rapist or murderer, still maintaining innocence (as, let's keep in mind, plenty of guilty felons do), demands a DNA test? Or claims that his or her confession was coerced? Or produces evidence suggesting that the eyewitness testimony that led to conviction was wrong? (About three-fourths of all DNA exonerations are cases that involved mistaken identification on the part of eyewitnesses.[41]) What if the defendant is not a monster, after all that hard work the legal team put in to convince themselves and the jury that he is? The response of prosecutors in Florida is typical. After more than 130 prisoners had been freed by DNA testing in the space of fifteen years, prosecutors decided they would respond by mounting a vigorous challenge to similar new cases. Convicted rapist Wilton Dedge had to sue the state to have the evidence in his case retested, over the fierce objections of prosecutors who said that the state's interest in finality and the victim's feelings should supersede concerns about Dedge's possible innocence.[42] Dedge was ultimately exonerated and released.

That finality and the victim's feelings should preclude justice seems an appalling argument by those we trust to provide justice, but that's the power of self-justification. Besides, wouldn't the victims feel better if the real perpetrators were caught and punished? DNA testing has freed hundreds of prisoners, and news accounts across the country often include a quote or two from the prosecutors who originally tried them. In Philadelphia, then district attorney Bruce L. Castor Jr. refused to accept the results of a DNA test that exonerated a man who had been in prison for fifteen years. When reporters asked him what scientific basis he had for rejecting the test, he replied, "I have no scientific basis. I know because I trust my detective and my tape-recorded confession."[43]

How do we know that this casual dismissal of DNA testing is a sign of self-justification and not simply an honest assessment of the evidence? It's like the horse-race study we described in chapter 1: Once you have placed your bets, you don't want to entertain any ideas that cast doubt on that decision. That is why prosecutors interpret the same evidence in one of two ways, depending on when it is discovered. Early in an investigation, the police use DNA to confirm a suspect's guilt or rule the person out. But when DNA tests are conducted after a defendant has been indicted and convicted, the prosecutors typically dismiss DNA results as irrelevant, not important enough to re-open the case. Texas prosecutor Michael McDougal said that the fact that the DNA found in a young rape-murder victim did not match that of Roy Criner, the man convicted of the crime, did not mean Criner was innocent. "It means that the sperm found in her was not his," he said. "It doesn't mean he didn't rape her, doesn't mean he didn't kill her."[44]

Technically, McDougal is right; Criner could have raped the woman in Texas and ejaculated somewhere else — Arkansas, per-haps. But DNA evidence should be used the same way whenever it turns up; it is the need for self-justification that prevents most prosecutors from being able to do that. Defense attorney Peter J. Neufeld says that in his experience, reinterpreting the evidence to justify the original verdict is extremely common among pros-ecutors and judges. During a trial, the prosecutor's theory is that one person alone, the defendant, seized and raped the vic-tim. If, after the defendant is convicted, DNA testing excludes him as the perpetrator, prosecutors miraculously come up with other theories. Our own favorite is what Neufeld calls the "unin-dicted co-ejaculator" theory: The convicted defendant held the woman down while a mysterious second man actually commit-ted the rape. Or the victim was lying there helpless, and a male predator "comes along and sees an opportunity and takes it," as one prosecutor claimed.[45] Or the defendant wore a condom, and the victim had consensual sex with someone else shortly before she was raped. (When Roy Criner's case was sent to the Texas Court of Criminal Appeals, Chief Judge Sharon Keller ruled that DNA "showing the sperm was not that of a man convicted of rape was not determinative because he might have worn a condom.") If the victim protests that she has not had intercourse in the previous three days, prosecutors advance the theory — again, after the trial — that she is lying: She doesn't want to admit that she had illicit sex because her husband or boyfriend will be angry.

In Rock Hill, South Carolina, Billy Wayne Cope, a socially isolated white man, was coerced into confessing to the rape and murder of his twelve-year-old daughter, although not a single piece of physical evidence linked him to the crime. Cope, a born-again Christian, thought his daughter might have died in the Rapture and then asked plaintively if he could have killed her in his sleep. When the DNA tests came back, they clearly identified the culprit as a black career criminal and serial rapist named James Sanders, a man who had always acted alone. Yet the men were tried together. "The only logical explanation," said the prosecutor, "is that Billy Cope served up his daughter for his and James Sanders's own perverse pleasures and took her life. They did it together. There is no other reasonable explanation."[46] How about the "reasonable explanation" that Billy Wayne Cope was innocent? In 2014, the South Carolina Supreme Court upheld Cope's sentence of life in prison. He died there three years later, maintaining his innocence to the end.

Self-justifications like these create a double tragedy: They keep innocent people in prison and allow the guilty to remain free. The same DNA that exonerates an innocent person can be used to identify the guilty one, but this does not happen nearly as often as it should.[47] The police and prosecutors usually prefer to close the books on the case completely, as if to obliterate its silent accusation of the mistake they made.

Jumping to Convictions

If the system can't function fairly, if the system can't correct its own mistakes and admit that it makes mistakes
and give people an opportunity to [correct] them, then
the system is broken.

　— *Michael Charlton, the appellate lawyer who*
　represented Roy Criner

All citizens have the right to a criminal justice system that con-
victs the guilty, protects the innocent, and corrects its mistakes
with alacrity. Legal scholars and social scientists have suggested
various constitutional remedies and important piecemeal improvements to reduce the risk of false confessions, unreliable eyewitness testimony, police "testilying," and so forth.[48] But from
our vantage point, the greatest impediment to admitting and
correcting mistakes in the criminal justice system is that most of
its members reduce dissonance by denying that there is a problem. "Our system has to create this aura of close to perfection,
of certainty that we don't convict innocent people," said former
prosecutor Bennett Gershman.[49] The benefit of this certainty
to police officers, detectives, and prosecutors is that they do not
spend sleepless nights worrying that they might have put an innocent person in prison. But a few sleepless nights are called for.
Doubt is not the enemy of justice; overconfidence is.

　Currently, the professional training of most police officers,
detectives, judges, and attorneys includes almost no informa-

tion about their own cognitive biases; how to correct for them, as much as possible; and how to manage the dissonance they will feel when their beliefs meet disconfirming evidence. On the contrary, much of what they learn about psychology comes from self-proclaimed experts with no training in psychological science and who, as we saw, do not teach them to be more accurate in their judgments, merely more confident that they are accurate: "An innocent person would never confess." "I saw it with my own eyes; therefore I'm right." "I can tell when someone is lying; I've been doing this for years." Yet that kind of certainty is the hallmark of pseudoscience. True scientists speak in the careful language of probability — "Innocent people most certainly can be induced to confess, under particular conditions; let me explain why I think this individual's confession is likely to have been coerced" — which is why scientists' testimony is often exasperating. Many judges, jurors, and police officers prefer certainties to science. Law professor D. Michael Risinger and attorney Jeffrey L. Loop have lamented "the general failure of the law to reflect virtually any of the insights of modern research on the characteristics of human perception, cognition, memory, inference or decision under uncertainty, either in the structure of the rules of evidence themselves, or the ways in which judges are trained or instructed to administer them."[50]

Yet training that promotes the certainties of pseudoscience rather than a humbling appreciation of our cognitive biases and blind spots increases the chances of wrongful convictions in two ways. First, it encourages law enforcement officials to jump to conclusions too quickly. A police officer decides that a suspect is the guilty party and then closes the door to other possibilities. A district attorney decides impulsively to prosecute a case, especially a sensational one, without having all the evidence; she announces her decision to the media and then finds it difficult to back down when subsequent evidence proves shaky. Second, once a case is prosecuted and a conviction won, officials will be motivated to reject any subsequent evidence of the defendant's innocence.

The antidote to these all-too-human mistakes is to ensure that in police academies and law schools, students learn about cognitive dissonance and their own vulnerability to self-justification. Mark Godsey, a law professor and former prosecutor who became an attorney for the wrongfully convicted, certainly did. In his book *Blind Justice,* he described his shock when one of the police department's snitches told him he had never confessed to an earlier crime he'd been convicted of; the detective had fabricated the confession. "Although to this day I don't know whether the informant was telling the truth about the allegedly fabricated confession," Godsey wrote, "I do know that I blew off his allegation because of cognitive dissonance. I swept it under a mental rug because it didn't coincide with my beliefs about the system. . . . We believe a snitch when he gives us information that helps us send someone to prison for life, but when he challenges our basic beliefs about the system, his allegations are promptly denied as nonsense without a closer look."[51]

That is precisely the kind of reasoning skill that all participants in the criminal justice system need to acquire. They must learn to look for the statistically likely suspect (a jealous boyfriend) without closing their minds to the statistically less likely suspect if that is where some evidence leads. They need to learn that even if they are confident that they can tell if a suspect is lying, they could be wrong. They need to learn how and why innocent people can be induced to confess to a crime they did not commit and how to distinguish confessions that are likely true from those that have been coerced.[52] They need to learn that the popular method of profiling, that beloved staple of the FBI and TV shows, carries significant risks of error because of the confirmation bias: when investigators start looking for elements of a crime that match a suspect's profile, they also start overlooking elements that do not match. In short, investigators need to learn to find another tree once they realize they are barking up the wrong one.

Law professor Andrew McClurg would go further in the training of police. He has long advocated the application of cognitive dissonance principles to keep highly motivated rookies from taking that first step down the pyramid in a dishonest direction; the centerpiece of his plan is to call on their own self-concept as good guys fighting crime and violence. He proposes a program of integrity training in dealing with ethical dilemmas, in which cadets would be instilled with the values of telling the truth and doing the right thing as a central part of their emerging professional identity. (Currently, in most jurisdictions, police trainees get one evening or a couple of hours on dealing with ethical problems.) Because such values are quickly trumped on the job by competing moral codes — "You don't rat on a fellow officer"; "In the real world, the only sure way to get a conviction is to fudge the truth" — McClurg proposes that rookies be partnered with experienced, ethical mentors who, in the manner of Alcoholics Anonymous sponsors, would help rookies maintain their honesty commitment. "The only hope of substantially re-ducing police lying is a preventative approach aimed at keeping good cops from turning bad," he argues. Cognitive dissonance theory offers "a potent, inexpensive, and inexhaustible tool for accomplishing this goal: the officer's own self-concept."[53]

Because no one, no matter how well trained or well inten-tioned, is completely immune to the confirmation bias and his or her own cognitive blind spots, the leading social scientists who have studied wrongful conviction are unanimous in recommend-ing safeguards, such as the electronic recording of all interviews. As of 2019, only twenty-six states plus the District of Columbia require the police to electronically record interrogations in some or all felony crimes, although only five states stipulate a "prefer-ence" for audiovisual recording.[54] Police and prosecutors have long resisted this requirement, fearing, perhaps, the embarrass-ing, dissonance-generating revelations it might create.

Ralph Lacer, one of the interrogators of Bradley Page, justi-fied the police position against videos on the grounds that a re-cording "is inhibiting" and makes it "hard to get at the truth."[55] Suppose, he complained, the interview goes on for ten hours. The defense attorney will make the jury listen to all ten hours instead of just the fifteen-minute confession, and the jury will be confused and overwhelmed. Yet in the Page case, the pros-ecution's argument rested heavily on a segment of the audio-taped interview that was missing. Lacer admitted that he had turned off the cassette player just before he said the words that convinced Page to confess. According to Page, during that miss-ing segment, Lacer had asked him to imagine how he *might* have killed his girlfriend. (This is another maneuver recommended by the creators of the Reid Technique.) Page thought he was being asked to construct an imaginary scenario to help the police; he was stunned when Lacer used it as a legitimate confession. The jury did not hear the full context — the question that elicited the alleged confession.

In fact, in jurisdictions that do record interrogations, law enforcement has come to favor the practice. The Center on Wrongful Convictions surveyed 238 law enforcement agencies that currently record all interrogations of felony suspects and found that almost all of the officers were enthusiastic. Videos made with a camera angle that includes both interviewer and interviewee eliminate the problem of suspects changing their stories, satisfy jurors that the confession was obtained honestly, and permit independent experts to assess the techniques that were used and determine whether any of them were deceptive or coercive.[56]

Canada and Great Britain are implementing these reforms and instituting other procedures to minimize the chances of wrongful convictions. One alternative to the coercive Reid Technique is the PEACE model (for Preparation and Planning, Engage and Explain, Account, Closure, Evaluate), which is used throughout the UK; variations of it are being developed in the United States. Users of the PEACE approach and similar methods do not presume the suspect's guilt, and the interview is not overtly confrontational; the police are not allowed to rely on bluffing and lying. The interview is seen as a way to get information; the interviewer asks open-ended questions in different versions, seeking the whole story and considering many possibilities. The assumption is that suspects who are lying have a "cognitive load" — caused by trying to remember false details — that is difficult to maintain.[57]

In the United States, the many exonerations due to DNA testing are also slowly bringing legal changes: improved oversight of crime labs, tougher standards for eyewitness identification, giving inmates (varying) degrees of access to DNA evidence, and, in a few states, the creation of commissions to expedite cases of wrongful conviction and find remedies. These commissions are almost invariably made up of prosecutors who were not themselves involved in the original cases and thus have no dissonance to reduce. On taking office in Brooklyn, DA Kenneth Thompson was alarmed to discover more than one hundred claims of wrongful convictions; he immediately created a conviction-review unit with ten prosecutors who did nothing but focus on those cases. So did Craig Watkins, DA in Dallas, who founded the Conviction Integrity Unit in 2007; it was expanded in 2017. Among its procedures is the systematic testing of DNA samples that had been overlooked in convictions; several dozen prisoners have since been exonerated. Texas has also passed a bill known as the junk-science statute; it permits a defendant to bring a writ of habeas corpus on the basis of new scientific evidence that indicates that the evidence that had been used to convict was false, misleading, or inaccurately applied. California likewise has passed a bill that permits convicted persons to contest expert testimony that was presented against them at trial, either because those experts subsequently repudiated their testimony or because they relied on methods or findings that were later shown to be faulty. These reforms are essential and long overdue.

But according to legal scholars and social scientists Debo-

rah Davis and Richard Leo, American law enforcement remains
steeped in its traditions, including adherence to the Reid Technique and similar procedures, maintaining "near absolute denial" that these techniques can and do produce false confessions
and wrongful convictions.[58] The fourth and fifth editions of the
Reid manual do include a disdainful nod to the problem of false
confessions, presumably to reassure readers that the authors are
aware of the cases that have made the news. But it is a grudg-
ing, selective review of the evidence that contains many errors
and doesn't acknowledge the extent of the problem, let alone the
Reid Technique's role in creating it. As Richard Leo puts it, the
manual pretends to be sensitive to the problem of false confessions without changing anything about the underlying method
itself, and people who go to their training classes report that in-
structors barely mention false confessions.

Two social scientists who have reviewed the research on the
Reid Technique observed that "most detectives are intelligent,
conscientious, and committed to a just outcome." But they conduct their interrogations according to the training they have received, almost invariably the Reid method, whose "vast edifice
of pseudoscience, misinformation, self-delusion and outright deceit does not advance the objectives of the criminal justice system. In the 1950s, it was heralded as a vast improvement over
the barbaric methods it replaced. Such justification stopped
being applicable decades ago."[59] Eric Shepherd, one of the psychologists involved in creating PEACE, agrees. "I think the Reid
Technique was a child of its time," he told the *New Yorker* reporter Douglas Starr. "What you see now is a rear-guard action
to defend the indefensible."[60]

The American criminal justice system's unwillingness to admit fallibility compounds the injustices it creates. Many states do absolutely nothing for people who have been exonerated. They provide no compensation for the many years of life and earnings lost. They do not even offer an official apology. Cruelly, they often do not expunge the exonerated person's record, making it difficult for the person to get an apartment or a job.

From the viewpoint of dissonance theory, we can see why the victims of wrongful convictions are treated so harshly. That harshness is in direct proportion to the system's inflexibility. If you know that errors are inevitable, you will not be surprised when they happen and you will have contingencies in place to remedy them. But if you refuse to admit to yourself or the world that mistakes do happen, then the exoneration of those who have been wrongfully imprisoned is stark, humiliating evidence of how wrong you are. Apologize to them? Give them money? Don't be absurd. They got off on a technicality. Oh, the technicality was DNA? Well, they were guilty of something else.

• • •

Once in a while, a man or woman of integrity rises above the common impulse to sacrifice truth in the service of self-justification: a police officer blows the whistle on corruption; a detective reopens a case that was apparently solved; a district attorney owns up to a miscarriage of justice. Indiana attorney Thomas Vanes was a prosecutor for thirteen years. "I was not bashful then in seeking the death penalty," he wrote.[61] "When criminals are guilty, they deserve to be punished." But Vanes learned that mistakes were made and that he had made them too.

I learned that a man named Larry Mayes, whom I had prosecuted and convicted, had served more than 20 years for a rape he did not commit. How do we know? DNA testing . . . Two decades later, when he requested a DNA retest on that rape kit, I assisted in tracking down the old evidence, convinced that the current tests would put to rest his long-standing claim of innocence. But he was right, and I was wrong.

Hard facts trumped opinion and belief, as they should. It was a sobering lesson, and none of the easy-to-reach rationalizations (just doing my job, it was the jurors who convicted him, the appellate courts had upheld the conviction) completely lessen the sense of responsibility — moral, if not legal — that comes with the conviction of an innocent man.

6

Love's Assassin:
Self-Justification in Marriage

Love . . . is the extremely difficult realization that something other than oneself is real.
— *Iris Murdoch, novelist*

When William Butler Yeats got married in 1917, his father wrote
him a warm letter of congratulations. "I think it will help you in
your poetic development," he said. "No one really knows human
nature, men as well as women, who has not lived in the bondage
of marriage, that is to say, the enforced study of a fellow creature."[1]
Married partners are forced to learn more about each other than
they ever expected (or perhaps wanted) to know. With no one
else, not even with our children or parents, do we learn so much
about another human being's adorable and irritating habits, ways
of handling frustrations and crises, and private, passionate desires. Yet, as John Butler Yeats knew, marriage also forces couples
to face themselves, to learn more about themselves and how they
behave with an intimate partner than they ever expected (or perhaps wanted) to know. No other relationship so profoundly tests

the extent of our own willingness to be flexible and forgiving, to learn and change — if we can resist the allure of self-justification.

Benjamin Franklin, who advised, "Keep your eyes wide open before marriage, and half shut afterward," understood the power of dissonance in relationships. Couples first justify their decision to be together and then their decision to stay together. When you buy a house, you start reducing dissonance immediately. You tell your friends the wonderful things you love about it (the view of the trees, the space, the original old windows) and minimize the things that are wrong with it (the view of the parking lot, the cramped kitchen, the drafty old windows). In this case, self-justification will keep you feeling happy about your beautiful new home. If, before you fell in love with it, a geologist had told you that the cliff above you was unstable and might give way at any moment, you would have welcomed the information and walked away, sad but not heartbroken. But once you have fallen in love with your house, spent more than you could really afford to buy it, and moved in with your unwilling cat, you have too much invested, emotionally and financially, to walk away easily. If, after you are in the house, someone tells you that the cliff above you is precarious, that same impulse to justify your decision may keep you there far too long. The people who live in houses along the beach in La Conchita, California, in the shadow of cliffs that have a habit of crashing down on them during heavy winter rains, live with constant dissonance, which they resolve by saying: "It won't happen again." This allows them to remain until it does happen again.

A relationship with a house is simpler than a relationship with another human being. For one thing, it's only one way. The house can't blame you for being a bad owner or for not keeping it clean, though it also can't give you a nice back rub after a hard day. Marriage, though, is the greatest two-way decision of most people's lives, and couples are enormously invested in making it work. A moderate amount of postwedding, eyes-half-shut dissonance reduction, in which partners emphasize the positive and overlook the negative, allows things to hum along in harmony. But the identical mechanism allows some people to remain in marriages that are the psychological equivalent of La Conchita, on the brink of constant disaster.

What do deliriously happy newlyweds have in common with unhappy couples who have remained together, in bitterness or weariness, for many years? An unwillingness to heed dissonant information. Many newlyweds, seeking confirmation that they have married the perfect person, overlook or dismiss any evidence that might be a warning sign of trouble or conflict ahead: "He goes into a sulk if I even chat with another man; how cute, it means he loves me." "She's so casual and relaxed about household matters; how charming, it means she'll make me less compulsive." Unhappy spouses who have long tolerated each other's cruelty, jealousy, or humiliation are also busy reducing dissonance. To avoid facing the devastating possibility that they have invested so many years, so much energy, so many arguments, in a failed effort to achieve even peaceful coexistence, they say something like "All marriages are like this. Nothing can be done about it, anyway. There are enough good things about it. Better to stay in a difficult marriage than be alone."

Self-justification doesn't care whether it reaps benefits or wreaks havoc. It keeps many marriages together (for better or worse) and tears others asunder (for better or worse). Couples start off blissfully optimistic, and over the years some will move in the direction of greater closeness and affection, others in the direction of greater distance and hostility. Some couples find in marriage a source of solace and joy, a place to replenish the soul, a relationship that allows them to flourish as individuals and as a couple. For others, marriage becomes a source of bickering and discord, a place of stagnation, a relationship that squashes their individuality and dissipates their bond. Our goal in this chapter is not to imply that all relationships can and should be saved, but rather to show how self-justification contributes to these two different outcomes.

Some couples separate because of a cataclysmic revelation, or ongoing violence that one partner can no longer tolerate or ignore. But the vast majority of couples who drift apart do so slowly, over time, in a snowballing pattern of blame and self-justification. Each partner focuses on what the other one is doing wrong while justifying his or her own preferences, attitudes, and ways of doing things. Each side's intransigence, in turn, makes the other side even more determined not to budge. Before the couple realize it, they have taken up polarized positions, each feeling right and righteous. Self-justification will then cause their hearts to harden against the entreaties of empathy.

• • •

To show how this process works, let's consider the marriage of Debra and Frank, taken from Andrew Christensen and Neil Jacobson's insightful book *Reconcilable Differences*.[2] Most people enjoy her-version/his-version accounts of a marriage (except when it's their own), shrugging their shoulders and concluding that there are two sides to every story. We think there's more to it than that.

This is Debra's version of their marital problems:

[Frank] just plods through life, always taking care of business, preoccupied with getting his work done but never showing much excitement or pain. He says his style shows how emotionally stable he is. I say it just shows he's passive and bored. In many ways I'm just the opposite: I have a lot of ups and downs. But most of the time I'm energetic, optimistic, spontaneous. Of course I get upset, angry, and frustrated sometimes. He says this range of feeling shows I'm emotionally immature, that "I have a lot of growing up to do." I think it just shows I'm human.

I remember one incident that kind of sums up the way I see Frank. We went out to dinner with a charming couple who had just moved to town. As the evening wore on, I became more and more aware of how wonderful their life was. They seemed genuinely in love with one another, even though they had been married longer than we have. No matter how much the man talked to us, he always kept in contact with his wife: touching her, or making eye contact with her, or including her in the conversation. And he used "we" a lot to refer to them. Watching them made me realize how little Frank and I touch, how rarely we look at each other, and how separately we participate in conversation. Anyway, I admit it, I was envious of this other couple. They seemed to have it all: loving family, beautiful home, leisure, luxury. What a contrast to Frank and me: struggling along, both working full-time jobs, trying to save money. I wouldn't mind that so much, if only we worked at it *together*. But we're so distant.

When we got home, I started expressing those feelings. I wanted to reevaluate our life — as a way of getting closer. Maybe we couldn't be as wealthy as these people, but there was no reason we couldn't have the closeness and warmth they had. As usual, Frank didn't want to talk about it. When he said he was tired and wanted to go to bed, I got angry. It was Friday night, and neither of us had to get up early the next day; the only thing keeping us from being together was his stubbornness. It made me mad. I was fed up with giving in to his need to sleep whenever I brought up an issue to discuss. I thought, Why can't he stay awake just for me sometimes?

I wouldn't let him sleep. When he turned off the lights, I turned them back on. When he rolled over to go to sleep, I kept talking. When he put a pillow over his head, I talked louder. He told me I was a baby. I told him he was insensitive. It escalated from there and got ugly. No violence but lots of words. He finally went to the guest bedroom, locked the door, and went to sleep. The next morning we were both worn out and distant. He criticized me for being so irrational. Which was probably true. I do get irrational when I get desperate. But I think he uses that accusation as a way of justifying himself. It's sort of like "If you're irrational, then I can dismiss all your complaints and I am blameless."

This is Frank's version:

Debra never seems to be satisfied. I'm never doing enough, never giving enough, never loving enough, never sharing enough. You name it, I don't do enough of it. Sometimes she gets me believing I really am a bad husband. I start feeling as though I've let her down, disappointed her, not met my obligations as a loving, supportive husband. But then I give myself a dose of reality. What have I done that's wrong? I'm an okay human being. People usually like me, respect me. I hold down a responsible job. I don't cheat on her or lie to her. I'm not a drunk or a gambler. I'm moderately attractive, and I'm a sensitive lover. I even make her laugh a lot. Yet I don't get an ounce of appreciation from her — just complaints that I'm not doing enough.

I'm not thrown by events the way Debra is. Her feelings are like a roller coaster: sometimes up, sometimes down. I can't live that way. A nice steady cruising speed is more my style. But I don't put Debra down for being the way she is. I'm basically a tolerant person. People, including spouses, come in all shapes and sizes. They aren't tailored to fit your particular needs. So I don't take offense at little annoyances; I don't feel compelled to talk about every difference or dislike; I don't feel every potential area of disagreement has to be explored in detail. I just let things ride. When I show that kind of tolerance, I expect my partner to do the same for me. When she doesn't, I get furious. When Debra picks at me about every detail that doesn't fit with her idea of what's right, I do react strongly. My cool disappears, and I explode.

I remember driving home with Debra after a night out with an attractive, impressive couple we had just met. On the way home I was wondering what kind of impression I'd made on them. I was tired that evening and not at my best. Sometimes I can be clever and funny in a small group, but not that night. Maybe I was trying too hard. Sometimes I have high standards for myself and get down on myself when I can't come up to them.

Debra interrupted my ruminations with a seemingly innocent question: "Did you notice how much in tune those two were with each other?" Now I know what's behind that kind of question — or at least where that kind of question will lead. It always leads right back to us, specifically to me. Eventually the point becomes "We aren't in tune with each other," which is code for "You're not in tune with me." I dread these conversations that chew over what's wrong with us as a couple, because the real question, which goes unstated in the civil conversations, but gets stated bluntly in the uncivil ones, is "What's wrong with Frank?" So I sidestepped the issue on this occasion by answering that they were a nice couple.

But Debra pushed it. She insisted on evaluating them in comparison to us. They had money and intimacy. We had neither. Maybe we couldn't be wealthy, but we could at least be intimate. Why couldn't we be intimate? Meaning: Why couldn't *I* be intimate? When we got home, I tried to defuse the tension by saying I was tired and suggesting that we go to bed. I *was* tired, and the last thing I wanted was one of these conversations. But Debra was relentless. She argued that there was no reason we couldn't stay up and discuss this. I proceeded with my bedtime routine, giving her the most minimal of responses. If she won't respect my feelings, why should I respect hers? She talked at me while I put on my pajamas and brushed my teeth; she wouldn't even let me alone in the bathroom. When I finally got into bed and turned off the light, she turned it back on. I rolled over to go to sleep, but she kept talking. You'd think she'd have gotten the message when I put the pillow over my head — but no, she pulled it off. At that point I lost it. I told her she was a baby, a crazy person — I don't remember everything I said. Finally, in desperation, I went to the guest bedroom and locked the door. I was too upset to go to sleep right away, and I didn't sleep at all. In the morning, I was still angry at her. I told her she was irrational. For once, she didn't have much to say.

Have you taken sides yet? Do you think this couple would be fine if she would only stop trying to get him to talk or if he would only stop hiding under the pillow, literally and figuratively? And what is their major problem — that they are temperamentally incompatible, that they don't understand each other, that they are angry?

All couples have differences. Even identical twins have differences. For Frank and Debra, like most couples, the differences are precisely why they fell in love: He thought she was terrific because she was sociable and outgoing, a perfect antidote to his reserve; she was drawn to his calmness and unflappability in a storm. All couples have conflicts too, small irritants that are amusing to observers but worthy of warfare to the participants (she wants dirty dishes washed immediately, and he lets them pile up so there's only one cleanup a day, or a week) or larger disagreements about money, sex, in-laws, or any of countless other issues. Differences need not cause rifts. But once there is a rift, the couple explains it as being an inevitable result of their differences.

Moreover, Frank and Debra actually understand their situation. They agree on everything that happened the night of their great blowup: on what set it off, on how they both behaved, on what each wanted from the other. They both agree that comparing themselves to the new couple made them feel unhappy and self-critical. They agree that she is more roller-coaster-y and he more placid, a gender complaint as common as ragweed in summer. They are clear about what they want from the relationship and what they feel they aren't getting. They even are very good, perhaps better than most, at understanding the other person's point of view.

Nor is this marriage deteriorating because Frank and Debra get angry at each other. Successful couples have conflicts and get angry just as unhappy couples do. But happy couples know how to manage their conflicts. If a problem is annoying them, they talk about and fix the problem, let it go, or learn to live with it.[3] Unhappy couples are pulled further apart by angry confrontations. When Frank and Debra get into a quarrel, they retreat to their familiar positions, brood, and stop listening to each other. If they do listen, they don't hear. Their attitude is: "Yeah, yeah,

I know how you feel about this, but I'm not going to change because I'm right."

To show what we think Frank and Debra's underlying problem is, let's rewrite the story of their trip home. Suppose that Frank had anticipated Debra's fears and concerns, which he knows very well by now, and expressed his genuine admiration for her sociability and ease with new people. Suppose, anticipating that she would compare their marriage unfavorably with this appealing couple's relationship, he said something like "You know, tonight I realized that even though we don't live in the luxury they do, I am awfully lucky to have you." Suppose that Frank had admitted candidly to Debra that being with this new couple made him feel "down on himself" about his participation that evening, a revelation that would have evoked her concern and sympathy. For her part, suppose that Debra had short-circuited her own self-pitying ruminations and paid attention to her husband's low mood, saying something like "Honey, you didn't seem to be up to par tonight. Are you feeling okay? Was it something about that couple you didn't like? Or were you just tired?" Suppose she, too, had been honest in expressing what she dislikes about herself, such as her envy of the other couple's affluence, instead of expressing what she dislikes about Frank. Suppose she had turned her attention to the qualities she does love about Frank. Hmmm, come to think of it, he's right about being a "sensitive lover."

From our standpoint, therefore, misunderstandings, conflicts, personality differences, and even angry quarrels are not the assassins of love; self-justification is. Frank and Debra's evening with the new couple might have ended very differently if both of them had not been so busy spinning their own self-justifications and blaming the other, and if they had thought about the other's feelings first. Each of them understands the other's point of view perfectly, but the need for self-justification is preventing them from accepting the other's position as legitimate. It is motivating each of them to see his or her own way as the better way, indeed the only reasonable way.

We are not referring here to the garden-variety kind of self-justification that we are all inclined to use when we make a mistake or disagree about relatively trivial matters, like who left the top off the salad dressing or who forgot to pay the water bill or whose memory of a favorite scene in an old movie is correct. In those circumstances, self-justification momentarily protects us from feeling clumsy, incompetent, or forgetful. The kind that can erode a marriage, however, reflects a more serious effort to protect not *what we did* but *who we are,* and it comes in two versions: "I'm right and you're wrong" and "Even if I'm wrong, too bad; that's the way I am." Frank and Debra are in trouble because they have begun to justify their fundamental self-concepts, the qualities about themselves that they value and do not wish to alter or that they believe are inherent in their nature. They are not saying to each other, "I'm right and you're wrong about that memory." They are saying, "I am the right kind of person and you are the wrong kind of person. And because you are the wrong kind of person, you cannot appreciate my virtues; foolishly, you even think some of my virtues are flaws."

Thus, Frank justifies himself by seeing his actions as those of a good, loyal, steady husband — that's who he is — and so he thinks their marriage would be fine if Debra quit pestering him to talk, if she would forgive his imperfections as he forgives hers. Notice his language: "What have I done that's wrong?" asks Frank. "I'm an okay human being." Frank justifies his unwillingness to discuss difficult or painful topics in the name of his "tolerance" and his ability to "just let things ride." For her part, Debra thinks her emotional expressiveness "just shows I'm human"—that's who she is—and that their marriage would be fine if Frank weren't so "passive and bored." Debra got it right when she observed that Frank justifies ignoring her demands to communicate by attributing them to her irrational nature. But she doesn't see that she is doing the same thing, that she justifies ignoring his wishes not to talk by attributing them to his stubborn nature.

Every marriage is a story, and like all stories, it is subject to its participants' distorted perceptions and memories that preserve the narrative as each side sees it. Frank and Debra are at a crucial decision point on the pyramid of their marriage, and the steps they take to resolve the dissonance between "I love this person" and "This person is doing some things that are driving me crazy" will enhance their love story or destroy it. They are going to have to decide how to answer some key questions about those crazy things their partner does: Are they due to an unchangeable personality flaw? Can I live with them? Are they grounds for divorce? Can we find a compromise? Could I—impossible thought that it is—learn something from my partner, maybe improve my own way of doing things? And they are going to have to decide how to think about their own way of doing things. Seeing as how they have lived with themselves their whole lives, "their own way" feels natural, inevitable. Self-justification is blocking each partner from asking: Could I be wrong? Could I be making a mistake? Could I change?

As Debra and Frank's problems accumulated, each developed an implicit theory of how the other person was wrecking the marriage. (These theories are called "implicit" because people are often unaware that they hold them.) Debra's implicit theory is that Frank is socially awkward and passive; his theory is that Debra is insecure and cannot accept herself or him as they are. The trouble is that once people develop an implicit theory, the confirmation bias kicks in and they stop seeing evidence that doesn't fit it. As Frank and Debra's therapist observed, Debra now ignores or plays down all the times that Frank isn't awkward and passive with her or others — the times he's been funny and charming, the many times he has gone out of his way to be helpful. For his part, Frank now ignores or plays down evidence of Debra's psychological security, such as her persistence and optimism in the face of disappointment. "They each think the other is at fault," their therapists observed, "and thus they selectively remember parts of their life, focusing on those parts that support their own points of view."[4]

Our implicit theories of why we and other people behave as we do come in one of two versions. We can say it's because of something in the situation or environment: "The bank teller snapped at me because she is overworked today; there aren't enough tellers to handle these lines." Or we can say it's because something is wrong with the person: "That teller snapped at me because she is plain rude." When we explain our own behavior, self-justification allows us to flatter ourselves: We give ourselves credit for our good actions but let the situation excuse the bad ones. When we do something that hurts another, we rarely say, "I behaved this way because I am a cruel and heartless human being." We say, "I was provoked; anyone would do what I did"; or "I had no choice"; or "Yes, I said some awful things, but that wasn't *me* — it's because I was drunk." Yet when we do something generous, helpful, or brave, we don't say we did it because we were provoked or drunk or had no choice or because the guy on the phone guilt-induced us into donating to charity. We did it because we are generous and open-hearted.

Successful partners extend to each other the same self-forgiving ways of thinking we extend to ourselves: They forgive each other's missteps as being due to the situation but give each other credit for the thoughtful and loving things they do. If one partner does something thoughtless or is in a crabby mood, the other tends to write it off as a result of events that aren't the partner's fault: "Poor guy, he is under a lot of stress"; "I can understand why she snapped at me; she's been living with back pain for days." But if one does something especially nice, the other credits the partner's inherent good nature and sweet personality: "My honey brought me flowers for no reason at all," a wife might say; "he is the dearest guy."

While happy partners are giving each other the benefit of the doubt, unhappy partners are doing just the opposite.[5] If the partner does something nice, it's because of a temporary fluke or situational demands: "Yeah, he brought me flowers, but only because all the other guys in his office were buying flowers for their wives." If the partner does something thoughtless or annoying, though, it's because of the partner's personality flaws: "She snapped at me because she's a bitch." Frank doesn't say that Debra did a crazy *thing*, following him around the house demanding that he talk to her, and he doesn't say she acted that way because she was feeling frustrated that he would not talk to her; he calls her a crazy *person*. Debra doesn't say that Frank avoided talking after the dinner party because he was weary and didn't want to have a confrontation last thing at night; she says he is a passive *person*.

Implicit theories have powerful consequences because they affect, among other things, how couples argue, and even the very purpose of an argument. If a couple is arguing from the premise that each is a good person who did something wrong but fixable, or who did something blunder-headed because of momentary situational pressures, there is hope of correction and compromise. But, once again, unhappy couples invert this premise. Because each partner is expert at self-justification, each blames the other's unwillingness to change on personality flaws but excuses his or her own unwillingness to change as personality virtues. If they don't want to admit they were wrong or modify a habit that annoys or distresses their partner, they say, "I can't help it. It's natural to raise your voice when you're angry. That's the way I am." You can hear the self-justification in these words because, of course, they *can* help it. They help it every time they don't raise their voice with a police officer, their employer, or a three-hundred-pound irritating stranger on the street.

The shouter who protests, "That's the way I am!" is, however, rarely inclined to extend the same self-forgiving justification to the partner. On the contrary, he or she is likely to turn it into an infuriating insult: "That's the way you are — you're just like your mother!" Generally, the remark does not refer to your mother's sublime baking skills or her talent at dancing the tango. It means that you are like your mother genetically and irredeemably; there's nothing you can do about it. And when people feel they can't do anything about it, they feel unjustly accused, as if they were being criticized for being too short or too freckled. Social psychologist June Tangney has found that being criticized for *who you are* rather than for *what you did* evokes a deep sense of shame and helplessness; it makes a person want to hide, disappear.[6] Because the shamed person has nowhere to go to escape the desolate feeling of humiliation, Tangney found, shamed spouses tend to strike back in anger: "You make me feel that I did an awful thing because I'm reprehensible and incompetent. Since I don't think I am reprehensible and incompetent, you must be reprehensible to humiliate me this way."

By the time a couple's style of argument has escalated into shaming and blaming each other, the fundamental purpose of their quarrels has shifted. It is no longer an effort to solve a problem or even to get the other person to modify his or her behavior; it's just to wound, to insult, to score. That is why shaming leads to fierce, renewed efforts at self-justification, a refusal to compromise, and the most destructive emotion a relationship can evoke: contempt. In his groundbreaking study of more than seven hundred couples whom he followed over a period of years, psychologist John Gottman found that contempt—criticism laced with sarcasm, name calling, and mockery—is one of the strongest signs that a relationship is in free fall.[7] Gottman offered this example:

FRED: Did you pick up my dry cleaning?

INGRID (*mocking*): "Did you pick up my dry cleaning?" Pick up your own damn dry cleaning. What am I, your maid?

FRED: Hardly. If you were a maid, at least you'd know how to clean.

Contemptuous exchanges like this one are devastating because they destroy the one thing that self-justification is designed to protect: our feelings of self-worth, of being loved, of being a good and respected person. Contempt is the final revelation to the partner that "I don't value the 'who' that you are." We believe that contempt is a predictor of divorce not because it causes the wish to separate but because it reflects the couple's feelings of psychological separation. Contempt emerges only after years of squabbles and quarrels that keep resulting, as for Frank and Debra, in yet another unsuccessful effort to get the other person to behave differently. It is an indication that the partner is throwing in the towel, thinking, "There's no point hoping that you will ever change; you are just like your mother after all." Anger reflects the hope that a problem can be corrected. When it burns out, it leaves the ashes of resentment and contempt. And contempt is the handmaiden of hopelessness.

• • •

Which comes first, a couple's unhappiness with each other or their negative ways of thinking about each other? Am I unhappy with you because of your personality flaws, or does my belief that you have personality flaws (rather than forgivable quirks or external pressures) eventually make me unhappy with you? Obviously it works in both directions. But because most new partners do not start out in a mood of complaining and blaming, psychologists have been able to follow couples over time to see what sets

some of them, but not others, on a downward spiral. They have
learned that negative ways of thinking and blaming usually come
first and are unrelated to the couple's frequency of anger or either party's feelings of depression.[8] Happy and unhappy partners
simply think differently about each other's behavior, even when
they are responding to identical situations and actions.

That is why we think that self-justification is the prime suspect in the murder of a marriage. Each partner resolves the dissonance caused by conflicts and irritations by explaining the
spouse's behavior in a particular way. That explanation, in turn,
sets them on a path down the pyramid. Those who travel the
route of shame and blame will eventually begin rewriting the
story of their marriage. As they do, they seek further evidence to
justify their growing pessimistic or contemptuous views of each
other. They shift from minimizing negative aspects of the marriage to overemphasizing them, seeking any bit of supporting evidence to fit their new story. As the new story takes shape, with
husband and wife rehearsing it privately or with sympathetic
friends, the partners become blind to each other's good qualities,
the ones that initially caused them to fall in love.

The tipping point at which a couple starts rewriting their love
story, Gottman finds, is when the "magic ratio" dips below five to
one: Successful couples have a ratio of five times as many positive
interactions (such as expressions of love, affection, and humor) to
negative ones (such as expressions of annoyance and complaints).
It doesn't matter if the couple is emotionally volatile, quarreling
eleven times a day, or emotionally placid, quarreling once a decade; it is the ratio that matters. "Volatile couples may yell and
scream a lot, but they spend five times as much of their marriage

being loving and making up," Gottman found. "Quieter, avoidant couples may not display as much passion as the other types, but they display far less criticism and contempt as well — the ratio is still 5 to 1."[9] When the ratio is five to one or better, any dissonance that arises is generally reduced in a positive direction. Social psychologist Ayala Pines, in a study of burnout in marriage, reported how a happily married woman she called Ellen reduced the dissonance caused by her husband's failure to give her a birthday present. "I wish he would have given me something — anything — I told him that, like I am telling him all of my thoughts and feelings," Ellen said to Pines. "And as I was doing that I was thinking to myself how wonderful it is that I can express openly all of my feelings, even the negative ones . . . The left over negative feelings I just sent down with the water under the bridge."[10]

When the positive-negative ratio has shifted toward those negative feelings, however, couples resolve dissonance *caused by the same events* in a way that increases their alienation from one another. Pines reported how an unhappily married woman, Donna, reacted to the same problem that upset Ellen: no birthday present from her husband. But whereas Ellen decided to accept that her husband was never going to become the Bill Gates of domestic giving, Donna interpreted her husband's behavior quite differently:

> One of the things that actually cemented my decision to divorce was my birthday, which is a symbolic day for me. I got a phone call at six o'clock in the morning from Europe, from a cousin, to wish me a happy birthday. Here is someone miles away who's taken the trouble. And he was

sitting there listening, and didn't wish me a happy birth-day . . . And I suddenly realized, you know, that here are all these people who do love me, and here's a person who doesn't appreciate me. He doesn't value me, he doesn't love me. If he did he wouldn't treat me the way he did. He would want to do something special for me.

It is entirely possible that Donna's husband doesn't love and appreciate her. And we don't have his side of the story about the birthday gift; perhaps he had tried giving her gifts for years but she never liked any of them. Presumably, though, most people don't decide to divorce because of a missing birthday present. Because Donna has decided that her husband's behavior is not only unmodifiable but intolerable, she now interprets every-thing he does as unmistakable evidence that "he doesn't value me, he doesn't love me." Donna actually took the confirmation bias further than most spouses do: She told Pines that when-ever her husband made her feel depressed and upset, she wrote it down in a "hate book." Her hate book gave her all the evidence she needed to justify her decision to divorce.

When the couple has hit this low point, they start revising their memories too. Now the incentive for both sides is not to send down the negative things "with the water under the bridge" but to encourage them to bubble up to the surface. Distortions of past events — or complete amnesia — kick in to confirm the couple's suspicion that they each married a complete stranger, and not a particularly appealing one either. Clinical psychologist Julie Gottman worked with an angry couple in therapy. When she asked, "How did the two of you meet?," the wife said, contemptuously, "At school, where I mistakenly thought he was smart."[11] In this twist of memory, she announces that she didn't make a mistake in choosing him; he made the mistake, by deceiving her about his intelligence.

"I have found that nothing foretells a marriage's future as accurately as how a couple retells their past," John Gottman observes.[12] Rewriting history begins even before a couple is aware the marriage is in danger. Gottman and his team conducted indepth interviews of fifty-six couples and were able to follow up with forty-seven of them three years later. At the time of the first interview, none of the couples had planned to separate, but the researchers were able to predict with 100 percent accuracy the seven couples who divorced. (Of the remaining forty couples, the researchers had predicted that thirty-seven would still be together, still an astonishing accuracy rate.) During the first interview, those seven couples had already begun recasting their history, offering a despondent story with confirming details to fit, telling Gottman they had married not because they were in love and couldn't bear to be apart but because marriage seemed "natural, the next step." The first year, the divorced couples now recalled, was full of letdowns and disappointments. "A lot of things went wrong but I don't remember what they were," said one soon-to-be-ex-husband. Happy couples, however, called the same difficulties "rough spots" and saw them proudly as challenges that they had survived, with humor and affection.

Thanks to the revisionist power of memory to justify our decisions, by the time many couples divorce, they can't remember why they married. It's as if they have had a nonsurgical lobotomy that excised the happy memories of how each partner once felt toward the other. Over and over we have heard people say, "I knew the week after the wedding I'd made a terrible mistake." "But why did you have three children and stay together for the next twenty-seven years?" "Oh, I don't know; I just felt obligated, I guess."

Obviously, some people do make the decision to separate as a result of a clear-eyed weighing of current benefits and problems; but for most, it's a decision fraught with historical revisionism and dissonance reduction. How do we know? Because even when the problems remain the same, the justifications change as soon as one or both parties decide to leave. As long as couples choose to stay in a relationship that is far from their ideal, they reduce dissonance in ways that support their decision: "It's not really that bad." "Most marriages are worse than mine — or certainly no better." "He forgot my birthday, but he does many other things that show me he loves me." "We have problems, but overall I love her." When one or both partners start thinking of divorce, however, their efforts to reduce dissonance will now justify the decision to leave: "This marriage really *is* that bad." "Most marriages are better than mine." "He forgot my birthday, and it means he doesn't love me." And the pitiless remark said by many a departing spouse after twenty or thirty years: "I never loved you."

The cruelty of that last particular lie is commensurate with the teller's need to justify his or her behavior. Spouses who leave a marriage because of clear external reasons — say, because a partner is physically or emotionally abusive — will feel no need for additional self-justification. Nor will those rare couples who part in complete amicability or who eventually restore warm feelings of friendship after the initial pain of separation. They feel no urgency to vilify their former partner or forget happier times, because they are able to say, "It didn't work out," "We just grew apart," or "We were so young when we married and didn't know better." But when the divorce is wrenching, momentous, and costly, and especially when one partner wants the separation and the other does not, both sides will feel an amalgam of painful emotions. In addition to the anger, anguish, hurt, and grief that almost invariably accompany divorce, these couples will also feel the pain of dissonance. That dissonance, and the way many people choose to resolve it, is one of the major reasons for postdivorce vindictiveness.

If you are the one being left, you may suffer the ego-crushing dissonance of "I'm a good person and I've been a terrific partner" and "My partner is leaving me. How could this be?" You could conclude that you're not as good a person as you thought or that you are a good person but you were a pretty bad partner, but few of us choose to reduce dissonance by plunging darts into our self-esteem. It's far easier to reduce dissonance by plunging darts into the partner, so to speak — say, by concluding that your partner is a difficult, selfish person, only you hadn't realized it fully until now.

If you are the one who is leaving, you also have dissonance to reduce, to justify the pain you are inflicting on someone you once loved. Because you are a good person, and a good person doesn't hurt another, your partner must have deserved your rejection, perhaps even more than you realized. Observers of divorcing couples are often baffled by what seems like unreasonable vindictiveness on the part of the person who initiated the separation; what they are observing is dissonance reduction in action. A friend of ours, lamenting her son's divorce, said: "I don't understand my daughter-in-law. She left my son for another man who adores her, but she won't marry him or work full-time just so that my son has to keep paying her alimony. My son has had to take a job he doesn't like, to afford her demands. Given that she's the one who left and that she has another relationship, the way she treats my son seems inexplicably cruel and vengeful." From the daughter-in-law's standpoint, however, her behavior toward her ex is perfectly justifiable. If he were such a good guy, she'd still be with him, wouldn't she? Therefore, since he hadn't been a good enough person to take a job he didn't like so she could live in the style she wanted, she'll make him do that *now.* Serves him right.

Divorce mediators, and anyone else who has tried to be helpful to warring friends in the throes of divorce, have seen this process up close. Mediators Donald Saposnek and Chip Rose described the "tendency of one spouse to cast the other in a vilified image, for example, 'He's a weak, violent drunk,' or, 'She's a two-faced, selfish, pathological liar who can't ever be trusted.' These intensely negative, polarized characterizations that high conflict divorcing couples make of each other become reified and immutable over time."[13] The reason they do is that once a couple starts reducing dissonance by taking the ego-preserving route of vilifying the former partner, they need to keep justifying their position. Thus they fight over every nickel and dime that one party is "entitled to" and the other "doesn't deserve," furiously denying or controlling custody matters and the ex's visitation rights because, look, the ex is a terrible person. Neither party pauses in mid-rant to consider if the ex's terribleness might be a result of the terrible situation, much less to consider if the

ex's terribleness might be a response to their own terrible behavior. Each action that one partner takes evokes a self-justified
retaliation from the other, and voilà, they are on a course of re-
ciprocal, escalating animosity. Each partner, having induced the
other to behave badly, uses that bad behavior both to justify his
or her own retaliation and to marshal support for the ex's inherently "evil" qualities.

By the time these couples seek mediation, they have slid
pretty far down the pyramid. Don Saposnek told us that in the
more than four thousand custody mediations he has done, "I
have *never* had one in which a parent has said, 'You know, I really think that she should get custody, since she really is the better parent and the kids are closer to her.' It is virtually always a
bilateral standoff of 'Why I am the better and more deserving
parent.' Not a single point of acknowledgment is ever given to
the other parent, and even when they freely admit their own acts
of retaliation, they always justify it: 'He deserved it, after what
he's done — breaking up our family!' The agreements they reach
are invariably some kind of compromise which each experiences
as 'giving up my position because I felt coerced, I'm exhausted
fighting, or I ran out of money for mediation ... even though I
know that I'm the better parent.'"

Dissonance theory would lead us to predict that it is the people who have the greatest initial ambivalence about their decision
to divorce — or who feel the greatest guilt over their unilateral
decision — who have the greatest urgency to justify their decision to leave. In turn, the bereft partner feels a desperate urgency
to justify any retaliation as payback for having been treated so

cruelly and unfairly. As both parties come up with confirming memories and all those horrible recent examples of the ex's bad behavior to support their new accounts, the ex turns completely villainous. Self-justification is the route by which ambivalence morphs into certainty, guilt into rage. The love story has become a hate book.

• • •

Our colleague Leonore Tiefer, a clinical psychologist, told us about a couple in their late thirties, married ten years, whom she saw in therapy. They could not make a decision about having children because each wanted to be sure before even raising the issue with the other. They could not make a decision about how to balance her demanding business career with their activities together, because she felt justified in working as much as she wanted. They could not resolve their quarrels over his drinking, because he felt justified in drinking as much as he wanted. Each had had an affair, which they justified as being a response to the other's.

Yet their normal, if difficult, problems were not what doomed this marriage; their obstinate self-justifications were. "They do not know what to give up in order to be a couple," says Tiefer. "They each want to do what they feel entitled to do, and they can't discuss the important issues that affect them as a pair. And as long as they stay mad at each other, they don't have to discuss those matters, because discussion might actually require them to compromise or consider the partner's point of view. They have a very difficult time with empathy, each one feeling completely confident that the other's behavior is less reasonable than their own. So they bring up old resentments to justify their current position and their unwillingness to change, or forgive."

In contrast, the couples who grow together over the years have figured out a way to live with a minimum of self-justification, which is another way of saying that they are able to put empathy for the partner ahead of defending their own territory. Successful, stable couples are able to listen to each other's criticisms, concerns, and suggestions undefensively. In our terms, they are able to yield, just enough, on the self-justifying excuse "That's the kind of person I am." They reduce the dissonance caused by small irritations by overlooking them, and they reduce the dissonance caused by their mistakes and major problems by solving them.

We interviewed several couples who have been together for many years, the kind of couples Frank and Debra admired, who by their own accounts have an unusually tight and affectionate marriage. We didn't ask them, "What is the secret of your long marriage?" because people rarely know the answer; they will say something banal or unhelpful, such as "We never went to bed angry" or "We share a love of golf." (Plenty of happy couples do go to bed angry because they would rather not have an argument when they are dead tired, and plenty of happy couples do not share hobbies and interests.) Instead, we asked these couples, in effect, how, over the years, they had reduced the dissonance between "I love this person" and "This person is doing something that is driving me crazy."

One especially illuminating answer came from a couple we will call Charlie and Maxine, who have been married more than forty years. Like all couples, they have many small differences that could easily flare into irritation, but they have come to accept most of them as facts of life, not worth sulking about. Charlie says, "I like to eat dinner at five; my wife likes to eat at eight; we compromise — we eat at five to eight." The important thing about this couple is how they handle the big problems. When they first fell in love, in their early twenties, Charlie was attracted to a quality of serenity in Maxine's soul that he found irresistible; she was, he said, an oasis in a tumultuous world. She was attracted to his passionate energy, which he brought to every activity, from planning the perfect vacation to writing the perfect sentence. But the passionate quality she enjoyed in him when it was attached to love, sex, travel, music, and movies was alarming to her when it was attached to anger. When he was angry, he would yell and pound the table, something no one in her family had ever done. Within a few months of their marriage, she told him, tearfully, that his anger was frightening her.

Charlie's first impulse was to justify himself. He didn't think that raising his voice was a desirable trait, exactly, but he saw it as one that was part of who he was, an aspect of his authenticity. "My father yelled and pounded tables," he said to her. "My grandfather yelled and pounded tables! It's my right! I can't do anything about it. It's what a man does. You want me to be like those wimpy guys who are always talking about their 'feelings'?" Once he stopped yelling and considered how his behavior was affecting Maxine, he realized that he could indeed modify his behavior, and, slowly and steadily, he reduced the frequency and intensity of his flare-ups. But Maxine, too, had to change; she had to stop justifying her belief that all forms of anger are dangerous and bad.

("In my family no one ever expressed anger. Therefore, that's the only right way to be.") When she did, she was able to learn to distinguish legitimate feelings of anger from unacceptable ways of expressing them, such as pounding tables, and for that matter from unconstructive ways of *not* expressing them, such as crying and retreating — her own "unchangeable" habit.

Over the years, a different problem emerged, one that had developed slowly, as it does for many couples who divide up tasks on the initial basis of who's better at them. The downside of Maxine's serenity was unassertiveness and a fear of confrontation; she would never dream of complaining about a bad meal or flawed merchandise. And so it always fell to Charlie to return the coffeepot that didn't work, call customer service with complaints, or deal face-to-face with the landlord who wouldn't fix the plumbing. "You're so much better at this than I am," she would say, and because he was, he would do it. Over time, however, Charlie grew tired of shouldering this responsibility and was becoming irritated by what he was now seeing as Maxine's passivity. "Why am I always the one handling these unpleasant confrontations?" he said to himself.

He was at a choice point. He could have let it slide, saying that's just the way she was, and continued to do all the dirty work. Instead, Charlie suggested that perhaps it was time for Maxine to learn how to be more assertive, a skill that would be useful to her in many contexts, not only in their marriage. Initially, Maxine responded by saying, "That's the way I am, and you knew it when you married me. Besides, no fair changing the rules after all these years." As they talked more, she was able to hear his concern without letting the jangle of self-justification get in the way. As soon as that happened, she could empathize with his feelings and understand why he thought the division of labor had become unfair. She realized that her options were not as limited as she had always assumed. She took an assertiveness-training course, diligently practiced what she learned there, got better at standing up for her rights, and before long was enjoying the satisfaction of speaking her mind in a way that usually got results. Charlie and Maxine made it clear that he did not turn into a lamb and she did not turn into a lion; personality, history, genetics, and temperament do put limitations on how much anyone can change.[14] But each of them moved. In this marriage, assertiveness and the constructive expression of anger are no longer polarized skills, his and hers.

In good marriages, a confrontation, difference of opinion, clashing habits, and even angry quarrels can bring the couple closer, by helping each partner learn something new and by forcing them to examine their assumptions about their abilities or limitations. It isn't always easy to do this. Letting go of the self-justifications that cover up our mistakes, that protect our desires to do things just the way we want to, and that minimize the hurts we inflict on those we love can be embarrassing and painful. Without self-justification, we might be left standing emotionally naked, unprotected, in a pool of regrets and losses.

• • •

No matter how painful it is to let go of self-justification, the result teaches us something deeply important about ourselves and can bring the peace of insight and self-acceptance. At the age of sixty-five, the feminist writer and activist Vivian Gornick wrote a dazzlingly honest essay about her lifelong efforts to balance work and love and lead a life based on exemplary egalitarian principles in both arenas. "I'd written often about living alone because I couldn't figure out why I *was* living alone," she wrote. For years her answer, the answer of so many in her generation, was sexism: Patriarchal men were forcing strong, independent women to choose between their careers and their relationships. That answer isn't wrong; sexism has sunk many marriages and shot holes through countless others that are barely afloat. But eventually Gornick realized that it was not the full answer. Looking back, without the comfort of her familiar self-justifications, she was able to see her own role in determining the course of her relationships, realizing "that much of my loneliness was self-inflicted, having more to do with my angry, self-divided personality than with sexism."[15]

"The reality was," she wrote, "that I was alone not because of my politics but because I did not know how to live in a decent way with another human being. In the name of equality I tormented every man who'd ever loved me until he left me: I called them on everything, never let anything go, held them up to accountability in ways that wearied us both. There was, of course, more than a grain of truth in everything I said, but those grains, no matter how numerous, need not have become the sandpile that crushed the life out of love."

7

Wounds, Rifts, and Wars

High-stomached are they both, and full of ire,
In rage deaf as the sea, hasty as fire.
    — *William Shakespeare,* Richard II

One year after he had confessed his affair, Jim felt there was no
letup in Karen's anger. Every conversation eventually turned to
the affair. She watched him like a hawk, and when he caught her
gaze, her expression was full of suspicion and pain. Couldn't she
realize that it had just been a small mistake on his part? He was
hardly the first person on the planet to make such a mistake. He
had been honest enough to admit the affair, after all, and strong
enough to end it. He had apologized and told her a thousand
times that he loved her and wanted the marriage to continue.
Couldn't she understand that? Couldn't she just focus on the
good parts of their marriage and get over this setback?

   Karen found Jim's attitude incredible. He seemed to expect
compliments for confessing the affair and ending it, rather than
criticism for having had the affair to begin with. Couldn't he un-

derstand that? Couldn't he just focus on her pain and distress and quit trying to justify himself? He had never even apologized either. Well, he said he was sorry, but that was pathetic. Why couldn't he give her a genuine, heartfelt apology? She didn't need him to prostrate himself; she just wanted him to know how she felt and make amends.

But Jim was finding it difficult to make the amends Karen wanted because of her intense anger, which made him feel like retaliating. The message he heard in her anger was "You have committed a horrible crime" and "You are less than human for doing what you did to me." He was deeply sorry that he had hurt her and he would give the world if he could only make her feel better, but he didn't think that he had committed a horrible crime or that he was inhuman, and the kind of groveling apology she seemed to want was not the kind he was prepared to give. So instead, he tried to convince her that the affair was not serious and that the other woman meant little to him. Karen, however, interpreted Jim's attempts to explain the affair as an effort to invalidate her feelings. The message she heard in his reaction was "You shouldn't be so upset; I didn't do anything bad." His efforts to explain himself made her angrier, and her anger made it more difficult for him to empathize with her suffering and respond to it.[1]

· · ·

The last battle in the terrible family war over the life and death of Terri Schiavo mesmerized millions of Americans. Terri's parents, Robert and Mary Schindler, had been fiercely fighting her husband, Michael Schiavo, over control of her life, or what remained of it. "It is almost beyond belief, given the sea of distance between them now, that Terri Schiavo's husband and parents once shared a home, a life, a goal," wrote one reporter. But it is not at all beyond belief to students of self-justification. At the start of Terri and Michael's marriage, the couple and her parents stood close together at the top of the pyramid. Michael called his in-laws Mom and Dad. The Schindlers paid the couple's rent in their early struggling years. When Terri Schiavo suffered massive brain damage in 1990, the Schindlers moved in with their daughter and son-in-law to jointly take care of her, and that is what they did for nearly three years. And then, the root of many rifts — money — was planted. In 1993, Michael Schiavo won a malpractice case against one of Terri's physicians and was awarded $750,000 for her care and $300,000 for the loss of his wife's companionship. A month later, husband and parents quarreled over the award. Michael Schiavo said it began when his father-in-law asked how much money he, Robert, would receive from the malpractice settlement. The Schindlers said the fight was about what kind of treatment the money should be spent on; the parents wanted intensive, experimental therapy and the husband wanted to give her only basic care.

The settlement was the first straw, forcing parents and husband to make a decision about how it should be spent and who deserved the money, because each side legitimately felt entitled to make the ultimate decisions about Terri's life and death. Accordingly, Michael Schiavo briefly blocked the Schindlers' access to his wife's medical records; they tried for a time to have him removed as her guardian. He was offended by what he saw as a crass effort by his father-in-law to claim some of the settlement money; they were offended by what they saw as his selfish motives to get rid of his wife.[2] By the time the country witnessed this family's final, furious confrontation, one inflamed by the media and opportunistic politicians, their reciprocally intransigent positions seemed utterly irrational and insoluble.

• • •

In January 1979, the shah of Iran, Mohammad Reza Pahlavi, faced with a growing public insurrection against him, fled Iran for safety in Egypt, and two weeks later the country welcomed the return of its new Islamic fundamentalist leader, Ayatollah Ruhollah Khomeini, whom the shah had sent into exile more than a decade earlier. In October, the Carter administration reluctantly permitted the shah to make a brief stopover in the United States on humanitarian grounds, for medical treatment for his cancer. Khomeini denounced the American government as the "Great Satan," urging Iranians to demonstrate against the United States and Israel, the "enemies of Islam." Thousands of them heeded his call and gathered outside the American embassy in Tehran. On November 4, several hundred Iranian students seized the main embassy building and took most of its occupants captive, fifty-two of whom remained as hostages for the next 444 days. The captors demanded that the shah be returned to Iran for trial, along with the billions of dollars they claimed the shah had stolen from the Iranian people. The Iran hostage crisis was the 9/11 of its day; according to one historian, it received more coverage on television and in the press than any other event had since World War II. Ted Koppel informed the nation of each day's (non)events in a new late-night show, *America Held Hostage,* which was so popular that when the crisis was

over it continued as *Nightline.* Americans were riveted by the
story, furious at the Iranians' actions and demands. So they were
mad at the shah; what the hell were they angry at us about?

· · ·

Thus far we have been talking about situations in which mistakes
were definitely made — memory distortions, wrongful convictions, misguided therapeutic practices. We move now to the far
more brambly territory of betrayals, rifts, and violent hostilities.
Our examples will range from family quarrels to the Crusades,
from routine meanness to systematic torture, from misdemeanors in marriage to the escalations of war. These conflicts between friends, cousins, and countries may differ profoundly in
cause and form, but they are woven together with the single, tenacious thread of self-justification. In pulling out that common
thread, we do not mean to overlook the complexity of the fabric
or imply that all garments are the same.

Sometimes both sides agree on who is to blame, as Jim and
Karen did; Jim did not try to shift the blame, as he might have
done by claiming that Karen drove him to have an affair by being
a bad wife. And sometimes it is all too certain who the guilty
party is even when the guilty party is busy denying it with a litany of excuses and self-justifications. Enslaved people are not
partly to blame for slavery, children do not provoke pedophiles,
women do not ask to be raped, the Jews did not bring the Holocaust on themselves.

We want to start, though, with a more common problem:
the many situations in which it isn't clear who is to blame, "who
started this," or even when this started. All families have tales to

tell of insults, unforgivable slights and wounds, and never-ending feuds: "She didn't come to my wedding, and she didn't even send a gift." "He stole my inheritance." "When my father was sick, my brother totally disappeared and I had to take care of him myself." In a rift, no one is going to admit that he or she lied or stole or cheated without provocation; only a bad person would do that, just as only a heartless child would abandon a parent in need. Therefore, each side justifies its own position by claiming that the other side is to blame; each is simply responding to the offense or provocation as any reasonable, moral person would do. "Yeah, you bet I didn't come to your wedding, and where were you seven years ago when I was going through that bad breakup and you vanished?" "Sure, I took some money and possessions from our parents' estate, but it wasn't stealing — you started this forty years ago when you got to go to college and I didn't." "Dad likes you better than me anyway, he was always so hypercritical of me, so it's right that you take care of him now."

In most rifts each side accuses the other of being inherently selfish, stubborn, mean, and aggressive, but the need for self-justification trumps personality traits. In all likelihood, the Schindlers and Michael Schiavo were not characteristically obstinate or irrational. Rather, their obstinate and irrational behavior in relation to each other was the result of twelve years of decisions (fight or yield on this one? Resist or compromise?), subsequent self-justifications, and further actions designed to reduce dissonance and ambivalence. Once they became entrapped by their choices, they could not find a way back. To justify their initial, understandable decision to keep their daughter alive, Terri's parents found themselves needing to justify their next decisions to keep her alive at all costs. Unable to accept the evidence that she was brain dead, Terri's parents justified their actions by accusing Michael of being a controlling husband, an adulterer, and possibly a murderer who wanted Terri to die because she had become a burden. To justify his equally understandable decision to let his wife die naturally, Michael, too, found himself on a course of action from which he could not turn back. To justify that course, he accused Terri's parents of being opportunistic media manipulators who were denying him the right to keep his promise to Terri that he would not let her live this way. The Schindlers were angry that Michael Schiavo would not listen to them or respect their religious beliefs. Michael Schiavo was angry that the Schindlers took the case to the courts and the public. Each side felt the other was behaving offensively; each felt profoundly betrayed by the other. Who started the final confrontation over control of Terri's death? Each says the other. What made it intractable? Self-justification.

When the Iranian students took those Americans hostage in 1979, the event seemed a meaningless act of aggression, a bolt that came out of the blue as far as the Americans were concerned; Americans saw themselves as having been attacked without provocation by a bunch of crazy Iranians. But to the Iranians, it was the Americans who started it, because American intelligence forces had aided in a coup in 1953 that unseated their charismatic, democratically elected leader, Mohammed Mossadegh, and installed the shah. Within a decade, many Iranians were growing resentful of the shah's accumulation of wealth and the westernizing influence of the United States. In 1963, the shah put down an Islamic fundamentalist uprising led by Khomeini and sent the cleric into exile. As opposition to the shah's government mounted, he allowed his secret police, SAVAK, to crack down on dissenters, fueling even greater anger.

When did the hostage crisis begin? When the United States supported the coup against Mossadegh? When it kept supplying the shah with arms? When it turned a blind eye to the cruelties committed by SAVAK? When it admitted the shah into the U.S. for medical treatment? Did it begin when the shah exiled Khomeini or when the ayatollah, after his triumphant return, saw a chance to consolidate his power by focusing the nation's frustrations on America? Did it begin during the protests at the embassy, when Iranian students allowed themselves to be Khomeini's political pawns? Most Iranians chose answers that justified their anger at the United States, and most Americans chose answers that justified their anger at Iran. Each side convinced itself that it was the injured party and consequently was entitled to retaliate. Who started the hostage crisis? Each says the other. What made it intractable? Self-justification.

Of all the stories that people construct to justify their lives, loves, and losses, the ones they weave to account for being the instigator or recipient of injustice or harm are the most compelling and have the most far-reaching consequences. In such cases, the hallmarks of self-justification transcend the specific antagonists (lovers, parents and children, friends, neighbors, or nations) and their specific quarrels (a sexual infidelity, a family inheritance, a betrayal of a confidence, a property line, or a military invasion). We have all done something that made others angry at us, and we have all been spurred to anger by what others have done to us. We all have, intentionally or unintentionally, hurt another person who will forever regard us as the villain, the betrayer, the scoundrel. And we have all felt the sting of being on the receiving end of an act of injustice, nursing a wound that never seems to fully heal. The remarkable thing about self-justification is that it allows us to shift from one role to the other and back again in the blink of an eye without applying what we have learned from one role to the other. Feeling like a victim of injustice in one situation does not make us less likely to commit an injustice against someone else, nor does it make us more sympathetic to victims. It's as if there is a brick wall between those two sets of experiences, blocking our ability to see the other side.

One of the reasons for that brick wall is that pain felt is always more intense than pain inflicted, even when the actual amount of pain is identical. The old joke — the other guy's broken leg is trivial; our broken fingernail is serious — turns out to be an accurate description of our neurological wiring. English neurologists paired people in a tit-for-tat experiment. Each pair was hooked up to a mechanism that exerted pressure on their index fingers, and the participants were instructed to apply the same force on their partner's finger that they had just felt. They could not do it fairly, although they tried hard. Every time one partner felt the pressure, he retaliated with considerably greater force, thinking he was giving what he had gotten. The researchers concluded that the escalation of pain is "a natural by-product of neural processing."[3] It helps explain why two boys who start out exchanging punches on the arm as a game soon find themselves in a furious fistfight. And it explains why two nations find themselves in a spiral of retaliation: "They didn't take an eye for an eye, they took an eye for a tooth. We must get even — let's take a leg." Each side justifies what it does as merely evening the score.

Social psychologist Roy Baumeister and his colleagues showed how smoothly self-justification works to minimize any bad feelings we have as doers of harm and to maximize any righteous feelings we have as victims.[4] They asked sixty-three people to provide autobiographical accounts of a "victim story," when they had been angered or hurt by someone else, and a "perpetrator story," a time when they had made someone else angry. They did not use the term *perpetrator* in its common criminal sense, to describe someone actually guilty of a crime or other wrongdoing, and in this section neither will we; we will use the word, as they do, to mean anyone who perpetrated an action that harmed or offended another.

From both perspectives, accounts involved the familiar litany of broken promises and commitments; violated rules, obligations, or expectations; sexual infidelity; betrayal of secrets; unfair treatment; lies; and conflicts over money and possessions. Notice that this was not a he-said/she-said study, the kind that marriage counselors and mediators present when they describe their cases; rather, it was a he-said-this-*and*-he-said-that study, in which everyone reported an experience of being on each side. The benefit of this method, the researchers explained, is that "it rules out explanations that treat victims and perpetrators as different kinds of people. Our procedures indicate how ordinary people define themselves as victims or as perpetrators — that is, how they construct narratives to make sense of their experiences in each of those roles." Again, personality differences have nothing to do with it. Sweet, kind people are as likely as crabby ones to be victims or perpetrators and to justify themselves accordingly.

When we construct narratives that "make sense," however,

we do so in a self-serving way. Perpetrators are motivated to reduce their moral culpability; victims are motivated to maximize
their moral blamelessness. Depending on which side of the wall
we are on, we systematically distort our memories and account
of the event to produce the maximum consonance between what
happened and how we see ourselves. By identifying these systematic distortions, the researchers showed how the two antagonists
misperceive and misunderstand each other's actions.

In their narratives, perpetrators drew on different ways to
reduce the dissonance caused by realizing they did something
wrong. The first, naturally, was to say they did nothing wrong at
all: "I lied to him, but it was only to protect his feelings." "Yeah, I
took that bracelet from my sister, but it was originally mine, anyway." Only a few perpetrators admitted that their behavior was
immoral or deliberately hurtful or malicious. Most said their offending behavior was justifiable, and some of them, the researchers added mildly, "were quite insistent about this." Most of the
perpetrators reported that what they did, at least in retrospect,
was reasonable; their actions might have been regrettable, but
they were understandable, given the circumstances.

The second strategy was to admit wrongdoing but excuse or
minimize it. "I know I shouldn't have had that one-night stand,
but in the great cosmos of things, what harm did it do?" "It might
have been wrong to take Mom's diamond bracelet when she was
ill, but she would have wanted me to have it. And besides, my
sisters got so much more than I did." More than two-thirds of
the perpetrators claimed external or mitigating circumstances
for what they did — "I was abused as a child myself"; "I've been
under a lot of stress lately" — but victims were disinclined to

grant their perpetrators these forgiving explanations. Nearly half of the perpetrators said they "couldn't help" what happened; they had simply acted impulsively, mindlessly. Others passed the buck, maintaining that the victim had provoked them or was otherwise partly responsible.

The third strategy, when the perpetrators were unequivocally caught and they could not deny or minimize responsibility, was to admit they had done something wrong and then try to get rid of the episode as fast as possible. Whether they accepted the blame or not, most perpetrators, eager to exorcise their dissonant feelings of guilt, bracketed the event off in time. They were far more likely than victims to describe the episode as an isolated incident that was now over and done with, that was not typical of them, that had no lasting negative consequences, and that certainly had no implications for the present. Many even told stories with happy endings that provided a reassuring sense of closure, along the lines of "everything is fine now, there was no damage to the relationship; in fact, today we are good friends."

For their part, the victims had a rather different take on the perpetrators' justifications, which might be summarized as "Oh, yeah? No damage? Good friends? Tell it to the Marines." Perpetrators may be motivated to get over the episode quickly and give it closure, but victims have long memories; an event that is trivial and forgettable to the former may be a source of lifelong rage to the latter. Only one of the sixty-three victim stories described the perpetrator as having been justified in behaving as he did, and none thought the perpetrators' actions "could not be helped." Accordingly, most victims reported lasting negative consequences of the rift or quarrel. More than half said it had seriously damaged the relationship. They reported continuing hostility, loss of trust, unresolved negative feelings, or even the end of the former friendship, which they apparently neglected to tell the perpetrator.

Moreover, whereas the perpetrators thought their behavior made sense at the time, many victims said they were unable to make sense of the perpetrators' intentions, even long after the event. "Why did he *do* that?" "What was she *thinking?*" The incomprehensibility of the perpetrator's motives is a central aspect of the victim identity and the victim story. "Not only did he do that terrible thing; he doesn't even understand that it *was* a terrible thing!" "Why can't she admit how mean she was to me?"

One reason he doesn't understand and she can't admit it is that perpetrators are preoccupied with justifying what they did, but another reason is that they really do not know how the victim feels. Many victims initially stifle their anger, nursing their wounds and brooding about what to do. They ruminate about their pain or grievances for months, sometimes for years, and sometimes for decades. One man we know told us that after eighteen years of marriage, his wife announced "out of the blue, at breakfast," that she wanted a divorce. "I tried to find out what I'd done wrong," he said, "and I told her I wanted to make amends, but there were eighteen years of dust balls under the bed." That wife brooded for eighteen years; the Iranians brooded for twenty-six years. By the time many victims get around to expressing their pain and anger, especially over events that the perpetrators have wrapped up and forgotten, perpetrators are baffled. No wonder most thought their victims' anger was an overreaction, though few victims felt that way. The victims are thinking, "Overreaction? But I thought about it for months before I spoke. I consider that an underreaction!"

Some victims justify their continued feelings of anger and their unwillingness to let it go because rage itself is retribution, a way to punish the offender, even when the offender wants to make peace, is long gone from the scene, or has died. In *Great Expectations,* Charles Dickens gave us the haunting figure of Miss Havisham, who, having been jilted on her wedding day, sacrifices the rest of her life to become a professional victim, clothed in self-righteous wrath and her yellowing bridal gown, raising her ward Estella to exact her revenge on men. Many victims are unable to resolve their feelings because they keep picking at the scab on the wound, asking themselves repeatedly, "How could such a bad thing have happened to me, a good person?" This is perhaps the most painful dissonance-arousing question that we confront in our lives. It is the reason for the countless books offering spiritual or psychological advice to help victims find closure — and consonance.

Whether it is Jim and Karen, Michael Schiavo and his in-laws, or the Iran hostage crisis, the gulf between perpetrators and victims can be seen in the way each side tells the same story. Perpetrators, whether individuals or nations, write versions of history in which their behavior was justified and provoked by the other side; their behavior was sensible and meaningful; if they made mistakes or went too far, at least everything turned out for the best in the long run; and it's all in the past now anyway. Victims tend to write accounts of the same history in which they describe the perpetrator's actions as arbitrary and meaningless, or else intentionally malicious and brutal; in which their own retaliation was impeccably appropriate and morally justified; and in which nothing turned out for the best. In fact, everything turned out for the worst, and we are still irritated about it.

Thus, Americans who live in the North and West learn about the Civil War as a matter of ancient history: "Our brave Union troops forced the South to abandon the ugly institution of slavery; we defeated the traitor Jefferson Davis, and the country remained united. (We'll just draw a veil over our own complicity as perpetrators and abettors of slavery; that was then.)" But most white Southerners tell a different story, one in which the Civil War is alive and kicking; then is *now:* "Our brave Confederate troops were victims of greedy, crude Northerners who defeated our noble leader Jefferson Davis, destroyed our cities and traditions, and are still trying to destroy our states' rights. There is nothing united about us Southerners and you damned Yankees; we'll keep flying our Confederate flag, thank you, that's *our* history." Slavery may be gone with the wind, but grudges aren't. That is why history is written by the victors, but it's victims who write the memoirs.

Who Started This?

One of the most eternally popular dissonance reducers, practiced by everyone from toddlers to tyrants, is "The other guy started it." Even Hitler said they started it, "they" being the victorious nations of World War I who humiliated Germany with the Treaty of Versailles and the Jewish "vermin" who were undermining Germany from within. The problem is, how far back do you want to go to show that the other guy started it? As our opening example of the Iran hostage crisis suggests, victims have long memories, and they can call on real or imagined episodes from the recent or distant past to justify their desire to retaliate now. In the centuries of war between Muslims and Christians, sometimes simmering and sometimes erupting, who are the perpetrators and who are the victims? There is no simple answer, but let's examine how each side has justified its actions.

After 9/11, George Bush announced that he was launching a crusade against terrorism, and most Americans welcomed the metaphor. In the West, *crusade* has positive connotations, associated with the good guys — Holy Cross's football team is the Crusaders and Batman and Robin are the Caped Crusaders. The actual historical Crusades in the Middle East began more than a thousand years ago and ended in the late thirteenth century; could anything be more over than that? Not to most Muslims, who were angered and alarmed by Bush's use of the term. For them, the Crusades created feelings of persecution and victimization that persist to the present. The First Crusade of 1095, during which Christians captured Muslim-controlled Jerusalem and mercilessly slaughtered almost all its inhabitants, might have occurred last month, it's that vivid in the collective memory.

The Crusades indeed gave European Christians license to massacre hundreds of thousands of Muslim "infidels." (Thousands of Jews were also slaughtered as the pilgrims marched through Europe to Jerusalem, which is why some Jewish historians call the Crusades "the first Holocaust.") From the West's current standpoint, the Crusades were unfortunate, but, like all wars, they produced benefits all around; for instance, the Crusades opened the door to cultural and trade agreements between the Christian West and the Muslim East. Some books have gone so far as to argue that Christians were merely defending themselves and their interests from the holy wars that had motivated the Muslim invasion of formerly Christian countries. The cover of Robert Spencer's *The Politically Incorrect Guide to Islam (and the Crusades)* states boldly: "The Crusades were defensive conflicts." So Christians were not the perpetrators that so many Muslims think they were. They were the victims.

Who *were* the victims? It depends on how many years, decades, and centuries you take into account. By the middle of the tenth century, more than a hundred years before the Crusades began, half the Christian world had been conquered by Muslim Arab armies: the city of Jerusalem and countries in which Christianity had been established for centuries, including Egypt, Sicily, Spain, and Turkey. In 1095, Pope Urban II called on the French aristocracy to wage holy war against all Muslims. A pilgrimage to regain Jerusalem would give European towns an opportunity to extend their trade routes; it would organize the newly affluent warrior aristocracy and mobilize the peasants into a unified force; and it would unite the Christian world, which had been split into Eastern and Roman factions. The pope assured his forces that killing a Muslim was an act of Christian penance. Anyone killed in battle, the pope promised, would bypass thousands of years of torture in purgatory and go directly to heaven. Does this incentive to generate martyrs who will die for your cause sound familiar? It has everything but the virgins.

The First Crusade was enormously successful in economic terms for European Christians; inevitably, it provoked the Muslims to organize a response. By the end of the twelfth century,

the Muslim general Saladin had recaptured Jerusalem and re-
taken almost every state the Crusaders had won. (Saladin signed
a peace treaty with King Richard I of England in 1192.) So the
Crusades, brutal and bloody as they were, were preceded and followed by Muslim conquests. Who started it?

Likewise, the intractable battles between Israelis and Palestinians have their own litany of original causes. On July 12, 2006,
Hezbollah militants kidnapped two Israeli reservists, Ehud
Goldwasser and Eldad Regev. Israel retaliated, sending rockets
into Hezbollah-controlled areas of Lebanon, killing many civilians. Historian Timothy Garton Ash, observing the subsequent
retaliations of both sides, wrote, "When and where did this war
begin?" Did it begin on July 12, or a month earlier, when Israeli
shells killed seven Palestinian civilians? The preceding January,
when Hamas won the Palestinian elections? In 1982, when Israel invaded Lebanon? In 1979, with the fundamentalist revolution in Iran? In 1948, with the creation of the State of Israel?
Garton Ash's own answer to "What started this?" is the virulent
European anti-Semitism of the nineteenth and twentieth centuries, which included Russian pogroms, French mobs screaming
"Down with Jews!" at the trial of Captain Alfred Dreyfus, and
the Holocaust. The "radical European rejection" of the Jews, he
writes, produced the driving forces of Zionism, Jewish emigra-
tion to Palestine, and the creation of the State of Israel:

> Even as we criticize the way the Israeli military is killing
> Lebanese civilians and U.N. monitors in the name of re-
> covering Goldwasser . . . we must remember that all of
> this almost certainly would not be happening if some

Europeans had not attempted, a few decades back, to remove everyone named Goldwasser from the face of Europe — if not the Earth.[5]

And Garton Ash was moving the start date back only a couple of centuries. Others would move it back a couple of millennia.

Once people commit themselves to an opinion about "who
started this?," whatever the "this" may be — a family quarrel or
an international conflict — they become less able to accept information that is dissonant with their positions. Once they have
decided who the perpetrator is and who the victim is, their ability to empathize with the other side is weakened, even destroyed.
How many arguments have you been in that sputtered out with
an unanswerable "But what about . . . ?" As soon as you describe
the atrocities that one side has committed, someone will protest:
"But what about the other side's atrocities?"

We can all understand why victims would want to retaliate.
But retaliation often makes the original perpetrators minimize
the severity and harm of their side's actions and claim the man-
tle of victim themselves, thereby setting in motion a cycle of op-
pression and revenge. "Every successful revolution," observed the
historian Barbara Tuchman, "puts on in time the robes of the ty-
rant it has deposed." Why not? The victors, former victims, feel
justified.

Perpetrators of Evil

The first shot I saw [from Abu Ghraib], of Specialist
Charles A. Graner and Pfc. Lynndie R. England flashing

thumbs up behind a pile of their naked victims, was so
jarring that for a few seconds I took it for a montage . . .
There was something familiar about that jaunty insou-
ciance, that unabashed triumph at having inflicted mis-
ery upon other humans. And then I remembered: the last
time I had seen that conjunction of elements was in photographs of lynchings.[6]
    — *Luc Sante, writer*

It may sometimes be hard to define good, but evil has
its unmistakable odor: Every child knows what pain is.
Therefore, each time we deliberately inflict pain on another, we know what we are doing. We are doing evil.[7]
    — *Amos Oz, novelist and social critic*

Did Charles Graner and Lynndie England believe they were
"doing evil" while they were deliberately inflicting pain and humiliation on their Iraqi prisoners and laughing at them? No, they
didn't, and that is why Amos Oz was wrong. Oz didn't reckon
with the power of self-justification: "We are good people. Therefore, if we deliberately inflict pain on another, the other must
have deserved it. Therefore, we are not doing evil, quite the contrary. We are doing good." Indeed, the small percentage of people who cannot or will not reduce dissonance this way pay a large
psychological price in guilt, anguish, anxiety, nightmares, and
sleepless nights, as we will discuss further in the next chapter.
The pain of living with horrors you have committed but cannot
morally accept is searing, which is why most people will reach
for any justification available to assuage the dissonance.

If good guys justify the bad things they do, bad guys persuade themselves they are the good guys. During his four-year trial for war crimes, crimes against humanity, and genocide, Slobodan Milosevic, the "Butcher of the Balkans," justified his policy of ethnic cleansing that resulted in the deaths of more than two hundred thousand Croats, Bosnian Muslims, and Albanians. He was not responsible for those deaths, he kept repeating at his trial; Serbs had been victims of Muslim propaganda. War is war; he was only responding to the aggression *they* perpetrated against the innocent Serbians. Riccardo Orizio interviewed seven other ruthless dictators, including Idi Amin, Jean-Claude "Baby Doc" Duvalier, Mira Markovic (the "Red Witch," Milosevic's wife), and Jean-Bédel Bokassa of the Central African Republic (known to his people as the Ogre of Berengo). Every one of them claimed that anything they did — torturing or murdering their opponents, blocking free elections, starving their citizens, looting their nation's wealth, launching genocidal wars — was done for the good of their country. The alternative, they said, was chaos, anarchy, and bloodshed. Far from seeing themselves as despots, they saw themselves as self-sacrificing patriots.[8] "The degree of cognitive dissonance involved in being a person who oppresses people out of love for them," wrote Louis Menand, "is summed up in a poster that Baby Doc Duvalier had put up in Haiti. It read, 'I should like to stand before the tribunal of history as the person who irreversibly founded democracy in Haiti.' And it was signed, 'Jean-Claude Duvalier, president-for-life.'"[9]

In the previous chapter, we saw on a smaller scale how divorcing couples typically justify the hurt they inflict on each other. In the horrifying calculus of self-deception, because our victims deserved what they got, we hate them even more than we did before we harmed them, which in turn makes us inflict even more pain on them. Experiments have confirmed this mechanism many times. In one experiment by Keith Davis and Edward Jones, students watched another student being interviewed and then, on instruction by the experimenters, had to report to the target student that they found him to be shallow, untrustworthy, and dull. As a result of making this rather nasty assessment, the participants succeeded in convincing themselves that the victim actually deserved their criticism, and they found him less appealing than they had before they hurt his feelings. Their change of heart occurred even though they knew that the other student had done nothing to merit their criticism and that they were simply following the experimenter's instructions.[10]

Are all victims alike in the eyes of the perpetrator? No; they differ in their degree of helplessness. Suppose you are a Marine in a hand-to-hand struggle with an armed enemy soldier. You kill him. Do you feel much dissonance? Probably not. The experience may be unpleasant, but it does not generate dissonance and needs no additional justification: "It was him or me . . . I killed an enemy . . . We are in this to win . . . I have no choice here." But now suppose that you are on a mission to firebomb a house that you were told contains enemy troops. You and your team destroy the place and then discover you have blown up a household of old men, children, and women. Under these circumstances, most soldiers will try to reduce the dissonance they feel about killing innocent civilians, and the leading way will be denigrating and dehumanizing their victims: "Stupid jerks, they shouldn't have been there . . . they were probably aiding the enemy . . . All those people are vermin, gooks, subhuman." Or, as General William Westmoreland famously said of the high number of civilian casualties during the Vietnam War, "The Oriental doesn't put the same high price on life as does a Westerner. Life is plentiful. Life is cheap in the Orient."[11]

Dissonance theory would therefore predict that when victims are armed and able to strike back, perpetrators will feel less need to reduce dissonance by belittling them than they do when their victims are helpless. In an experiment by Ellen Berscheid and her associates, participants were led to believe that they would be delivering a painful electric shock to another person as part of a test of learning. Half were told that later they would be reversing roles, so the victim would be in a position to retaliate. As predicted, the only participants who denigrated their victims were those who believed the victims were helpless and would not be able to respond in kind.[12] This was precisely the situation of the people who took part in Stanley Milgram's 1963 obedience experiment, described in chapter 1. Many of those who obeyed the experimenter's orders to deliver what they thought were dangerous amounts of shock to a learner justified their actions by blaming the victim. As Milgram himself put it, "Many subjects harshly devalue the victim *as a consequence* of acting against him. Such comments as, 'He was so stupid and stubborn he deserved to get shocked,' were common. Once having acted against the victim, these subjects found it necessary to view him as an unworthy individual, whose punishment was made inevitable by his own deficiencies of intellect and character."[13]

The implications of these studies are ominous, for they show that people do not perform acts of cruelty and come out un-

scathed. Success at dehumanizing the victim virtually guarantees a continuation or even an escalation of the cruelty: It sets
up an endless chain of violence, followed by self-justification (in
the form of dehumanizing and blaming the victim), followed by
still more violence and dehumanization. Combine self-justifying
perpetrators and victims who are helpless, and you have a recipe
for the escalation of brutality. This brutality is not confined to
brutes — that is, sadists or psychopaths. It can be, and usually is,
committed by ordinary individuals, people who have children
and lovers, "civilized" people who enjoy music and food and
making love and gossiping as much as anyone else. This is one of
the most thoroughly documented findings in social psychology,
but it is also the most difficult for many people to accept because
of the enormous dissonance it produces: "What can I possibly
have in common with perpetrators of murder and torture?" It is
much more reassuring to believe that they are evil and be done
with them.[14] We dare not let a glimmer of their humanity in the
door, because it might force us to face the haunting truth of car-
toonist Walt Kelly's great character Pogo, who famously said:
"We have met the enemy and he is us."

If the perpetrators are seen as one of us, however, many people will reduce dissonance by coming to their defense or minimizing the seriousness or illegality of their actions, anything that
makes their actions seem fundamentally different from what
the enemy does. They assume that only villains like Idi Amin or
Saddam Hussein would torture their enemies. But as John Conroy showed in *Unspeakable Acts, Ordinary People,* it is not only
interrogators in undemocratic countries who violate the Geneva
Conventions' prohibitions against "violence to life and person,

in particular murder of all kinds, mutilation, cruel treatment and torture . . . [and] outrages upon personal dignity, in particular, humiliating and degrading treatment." In his investigation of documented cases of abuse of prisoners, Conroy found that almost every military or police official he interviewed, whether British, South African, Israeli, or American, justified their practices by saying, in effect, "Our torture is never as severe and deadly as their torture":

> Bruce Moore-King [of South Africa] told me that when he administered electrical torture he never attacked the genitals, as torturers elsewhere are wont to do . . . Hugo Garcia told me the Argentine torturers were far worse than the Uruguayan. Omri Kochva assured me that the men of the Natal battalion had not descended to the level of the Americans in Vietnam . . . The British comforted themselves with the rationalization that their methods were nothing compared to the suffering created by the IRA. The Israelis regularly argue that their methods pale in comparison to the torture employed by Arab states.[15]

As for Americans: The photos of American soldiers humiliating and torturing terrorist suspects at Abu Ghraib prison in Iraq caused worldwide revulsion. Impartial investigations by the Red Cross, Amnesty International, and Human Rights Watch had revealed that American interrogators and their allies had been using sleep deprivation, prolonged isolation, waterboarding, sexual humiliation, induced hypothermia, beatings, and other cruel methods on terrorist suspects, not only at Abu Ghraib but also at Guantánamo Bay and at "black sites" in other countries. In 2014, a Senate Intelligence Committee report confirmed that the CIA's use of torture was more widespread and brutal than Congress or the public had been led to believe.[16]

How did the creators of CIA policy and those who carried it out reduce the dissonance caused by the information that the United States had been systematically violating the Geneva Conventions? The first way is to say that if we do it, it isn't torture. "We do not torture," said George W. Bush. "We use an alternative set of procedures." Dick Cheney's response to the 2014 Senate report, before he read it, was "It's full of crap." When Chuck Todd interviewed Cheney on *Meet the Press,* Todd persistently asked him to define torture. "There's this notion that somehow there's moral equivalence between what the terrorists and what we do," Cheney replied. "And that's absolutely not true. We were very careful to stop short of torture. The Senate has seen fit to label their report *torture.* But we worked hard to stay short of that definition."

Todd pressed him: Well, then, what is that definition? Isn't "rectal feeding" torture? Cheney replied, testily, "I've told you what meets the definition of torture. It's what nineteen guys armed with airline tickets and box cutters did to three thousand Americans on 9/11." Todd kept at it: How about Riyadh al-Najjar, who was handcuffed by his wrists to an overhead bar so that he could not lower his arms for twenty-two hours a day for two consecutive days — while wearing a diaper and denied access to a toilet? How about Abu Zubaydah, who was confined for eleven days and two hours in a coffin-size box that was twenty-one inches wide, two and a half feet deep, and two and a half feet high? Not torture, said Cheney; those were approved techniques. "Does the report plant any seed of doubt in you, though?" asked Todd. "Absolutely not," said Cheney.[17]

A second way to reduce dissonance is to say that if we do torture anyone, it's justified. The prisoners at Abu Ghraib deserved everything they got, said Senator James Inhofe of Oklahoma, because "they're murderers, they're terrorists, they're insurgents. Many of them probably have American blood on their hands." He seemed unaware that most of the prisoners had been picked up for arbitrary reasons or minor crimes and were never formally accused. Indeed, several military intelligence officers told the International Committee of the Red Cross that between 70 and 90 percent of the Iraqi detainees had been arrested by mistake.[18]

The universal justification for torture is the ticking-time-bomb excuse. As the columnist Charles Krauthammer put it, "A terrorist has planted a nuclear bomb in New York City. It will go off in one hour. A million people will die. You capture the terrorist. He knows where it is. He's not talking. Question: If you have the slightest belief that hanging this man by his thumbs will get you the information to save a million people, are you permitted to do it?" Yes, says Krauthammer, and not only are you permitted to, it's your moral duty.[19] You don't have time to call the Geneva Conventions people and ask them if it's okay; you will do whatever you can to get the terrorist to tell you the bomb's location.

When put that way, most Americans might set aside their moral qualms and decide that torturing one person is worth saving a million lives. The trouble with that line of reasoning is that its pragmatic justification doesn't hold up: tortured suspects will say *anything*. As one editorial put it, "Torture is a terrible way

to do the very thing that the administration uses to excuse it —
getting accurate information. Centuries of experience show that
people will tell their tormenters what they want to hear, whether
it's confessing to witchcraft in Salem, admitting to counterrevo-
lutionary tendencies in Soviet Russia or concocting stories about
Iraq and Al Qaeda."[20] Indeed, the Senate Intelligence report confirmed that *no* information gained from torturing detainees had
proved useful in capturing or killing any terrorist, including
Osama bin Laden. Worse, the "saving lives" excuse is used even
when there is no ticking and there is no bomb. Former Secretary
of State Condoleezza Rice, on a visit to Germany where she was
bombarded by protests from European leaders about the American use of torture on terrorist suspects held in secret jails, denied that any torture was being used. Then she added that her
critics should realize that interrogations of these suspects have
produced information that "stopped terrorist attacks and saved
innocent lives — in Europe as well as in the United States."[21] She
seemed unconcerned that these interrogations had also ruined
innocent lives. Rice reluctantly admitted that "mistakes were
made" when the United States abducted an innocent German
citizen on suspicions of terrorism and subjected him to harsh
and demeaning treatment for five months.

Once torture is justified in one case, it is easier to justify it
in others: "Let's torture not only this bastard who we are pretty
sure knows where the bomb is but this other bastard who *might*
know where the bomb is, and also this bastard who might have
some general information that could be useful in five years, and
also this other guy who might be a bastard only we aren't sure."
William Schulz, director of Amnesty International, observed

that according to credible Israeli, international, and Palestinian human rights organizations, Israelis used methods of interrogation from 1987 to 1993 that constituted torture. "While originally justified on the grounds of finding 'ticking bombs,'" he said, "the use of such methods of torture became routine."[22] A sergeant in the U.S. Army's Eighty-Second Airborne Division described how this process happened in treating Iraqi detainees:

> The "Murderous Maniacs" was what they called us at our camp . . . When [the detainees] came in, it was like a game. You know, how far could you make this guy go before he passes out or just collapses on you. From stress positions to keeping them up two days straight, depriving them of food, water, whatever . . . We were told by intel that these guys were bad, but sometimes they were wrong.[23]

"Sometimes they were wrong," the sergeant says, but nonetheless we treated them all the same way.

The debate about torture has properly focused on its legality, its morality, and its utility. As social psychologists, we want to add one additional concern: what torture does to the individual perpetrator and the ordinary citizens who go along with it. Most people want to believe that their government is working on their behalf, that it knows what it's doing, and that it's doing the right thing. Therefore, if the government decides that torture is necessary in the war against terrorism, most citizens, to avoid dissonance, will agree. Yet, over time, that is how the moral conscience of a nation deteriorates. Once people take that first small step off the pyramid in the direction of justifying abuse and torture, they are on their way to hardening their hearts and minds in ways that might never be undone. Uncritical patriotism, the kind that reduces the dissonance caused by information that their government — and especially their political party — has done something immoral and illegal, greases the slide down the pyramid.

We have watched this slide with sorrow and alarm. In December 2014, after the Senate Intelligence report appeared, a national Pew survey found that 51 percent of all Americans still agreed that the CIA's use of torture was "justified" and more than half still mistakenly believed that the CIA's interrogation methods had helped prevent terrorist attacks. The polls also showed that this issue, like so many others where the two parties once found common ground, had become a partisan matter: 76 percent of all Republicans said that the CIA's post-9/11 interrogation methods were justified, while just 37 percent of Democrats did.[24] But torture wasn't always a partisan issue; it was Ronald Reagan, after all, who signed the United Nations Convention Against Torture in 1988. And although Barack Obama banned the use of "enhanced interrogation techniques," he looked the other way on his predecessor's policy. "We tortured some folks," he said, softening the ugliness of the action by using that, dare we say, folksy term. He supported the CIA's efforts to censor parts of the report, and he declined to hold accountable any perpetrators of torture or the policies that permitted it.

"Without a mutual acknowledgment of mistakes made, and some form of accountability, another reversion to torture may be difficult to prevent," says political scientist Darius Rejali.

"Nothing predicts future behavior as much as past impunity."[25] Impunity, in turn, rewards self-justification, not only in the perpetrators but also in the nation that exonerates them.

Nevertheless, some politicians have resisted the temptation to justify the actions of the CIA, and one of them, Republican senator John McCain, was especially eloquent. "The truth is sometimes a hard pill to swallow," he said, "but the American people are entitled to it, nonetheless":

> They must know when the values that define our nation are intentionally disregarded by our security policies, even those policies that are conducted in secret. They must be able to make informed judgments about whether those policies and the personnel who supported them were justified in compromising our values; whether they served a greater good; or whether, as I believe, they stained our national honor, did much harm and little practical good. What were the policies? What was their purpose? Did they achieve it? Did they make us safer? Less safe? Or did they make no difference? What did they gain us? What did they cost us? The American people need the answers to these questions . . .
>
> [The use of torture] was shameful and unnecessary . . . But in the end, torture's failure to serve its intended purpose isn't the main reason to oppose its use. I have often said, and will always maintain, that this question isn't about our enemies; it's about us. It's about who we were, who we are and who we aspire to be. It's about how we represent ourselves to the world.[26]

Truth and Reconciliation

In our favorite version of an ancient Buddhist parable, several monks are returning to their monastery after a long pilgrimage. Over high mountains and across low valleys they trek, honoring their vow of silence outside the monastery. One day they come to a raging river where a beautiful young woman stands. She approaches the eldest monk and says, "Forgive me, Roshi, but would you be so kind as to carry me across the river? I cannot swim, and if I remain here or attempt to cross on my own, I shall surely perish." The old monk smiles at her warmly and says, "Of course I will help you." With that, he picks her up and carries her across the river. On the other side, he gently sets her down. She thanks him, departs, and the monks continue their wordless journey.

After five more days of arduous travel, the monks arrive at their monastery, and the moment they do, they turn on the elder in a fury. "How could you do that?" they admonish him. "You broke your vows! You not only spoke to that woman, you touched her! You not only touched her, you picked her up!"

The elder replies, "I only carried her across the river. You have been carrying her for five days."

The monks carried the woman in their hearts for days; some perpetrators and victims carry their burdens of guilt, grief, anger, and revenge for years. What does it take to set those burdens down? Anyone who has tried to intervene between warring couples or nations knows how painfully difficult it is for both sides to let go of self-justification, especially after years of fighting, defending their position, and moving farther down the pyramid away from compromise and common ground. Mediators and negotiators therefore have two challenges: to persuade perpetrators to acknowledge and atone for the harm they caused, and to persuade victims to relinquish the impulse for revenge while recognizing and sympathizing with the harm they have suffered.

In their work with married couples in which one partner had deeply hurt or betrayed the other, clinical psychologists Andrew Christensen and Neil Jacobson described three possible ways out of the emotional impasse. In the first, the perpetrator unilaterally puts aside his or her own feelings and, realizing that the victim's anger masks enormous suffering, responds to that suffering with genuine remorse and apology. In the second, the victim unilaterally lets go of his or her repeated, angry accusations — after all, the point has been made — and expresses pain rather than anger, a response that may make the perpetrator feel empathic and caring rather than defensive. "Either one of these actions, if taken unilaterally, is difficult and for many people impossible," Christensen and Jacobson say.[27] The third way, they suggest, is the hardest but most hopeful for a long-term resolution of the conflict: Both sides drop their self-justifications and agree on steps they can take together to move forward. If it is only the perpetrator who apologizes and tries to atone, it may not be done honestly or in a way that assuages and gives closure to the victim's suffering. But if it is only the victim who lets go and forgives, the perpetrator may have no incentive to change and therefore may continue behaving unfairly or callously.[28]

Christensen and Jacobson were speaking of two individuals in conflict. But their analysis, in our view, applies to group conflicts as well, where the third way is not merely the best way; it is the only way. In South Africa, the end of apartheid could easily have left a legacy of self-justifying rage on the part of the whites who supported the status quo and the privileges it conferred on them, and of self-justifying fury on the part of the blacks who had been its victims. It took the courage of a white man, Frederik de Klerk, and a black man, Nelson Mandela, to avert the blood-bath that follows in the wake of most revolutions and create the conditions that made it possible for their country to move forward as a democracy.

De Klerk, who had been elected president in 1989, knew that a violent revolution was all but inevitable. The fight against apartheid was escalating; sanctions imposed by other countries were having a significant impact on the nation's economy; supporters of the banned African National Congress were becoming increasingly violent, killing and torturing people whom they believed were collaborating with the white regime. De Klerk could have tightened the noose by instituting even more repressive policies in the desperate hope of preserving white power. Instead, he revoked the ban on the ANC and freed Mandela from the prison in which he had spent twenty-seven years. For his part, Mandela could have allowed his anger to consume him; he could have emerged from that prison with a determination to take revenge that many would have found entirely legitimate. Instead, he relinquished anger for the sake of the goal to which he had devoted his life. "If you want to make peace with your enemy, you have to work with your enemy," said Mandela. "Then he becomes your partner." In 1993, both men shared the Nobel Peace Prize, and the following year Mandela was elected president of South Africa.

Virtually the first act of the new democracy was the establishment of the Truth and Reconciliation Commission, chaired by Archbishop Desmond Tutu. (Three other commissions, on human rights violations, amnesty, and reparation and rehabilitation, were also created.) The goal of the TRC was to give victims of brutality a forum where their accounts would be heard and validated, where their dignity and sense of justice would be restored, and where they could express their grievances in front of the perpetrators themselves. In exchange for amnesty, the perpetrators had to admit to the harm they had done, including torture and murder. The commission emphasized the "need for understanding but not for vengeance, a need for reparation but not for retaliation, a need for *ubuntu* [humanity toward others] but not for victimization."

The goals of the TRC were inspiring, if not entirely honored in practice. The commission produced grumbling, mockery, protests, and anger. Many black victims of apartheid, such as the family of Stephen Biko, an activist who had been murdered in prison, were furious at the provisions of amnesty to the perpetrators. Many white perpetrators did not apologize with anything remotely like true feelings of remorse, and many white supporters of apartheid were not interested in listening to the broadcast confessions of their peers. South Africa has hardly become a paradise; it is still suffering from poverty and high crime rates. Yet the predicted eruption of violence did not occur. When psychologist Solomon Schimmel traveled there and interviewed people across the political and cultural spectrum for his book on victims of injustice and atrocities, he expected to hear them describe their rage and desire for revenge. But "what most impressed me overall," he reported, "was the remarkable lack of overt rancor and hatred between blacks and whites, and the concerted effort to create a society in which racial harmony and economic justice will prevail."[29]

• • •

Understanding without vengeance, reparation without retaliation, are possible only if we are willing to stop justifying our own position. Many years after the Vietnam War, veteran William Broyles Jr. traveled back to Vietnam to try to resolve his feelings about the horrors he had seen there and those he had committed. He went because, he said, he wanted to meet his former enemies "as people, not abstractions." In a small village that had been a Marine base camp, he met a woman who had been with the Viet Cong. As they talked, Broyles realized that her husband had been killed at exactly the time that he and his men had been patrolling. "My men and I might have killed your husband," he said. She looked at him steadily and said, "But that was during the war. The war is over now. Life goes on."[30] Later, Broyles reflected on his healing visit to Vietnam:

> I used to have nightmares. Since I've been back from that trip, I haven't had any. Maybe that sounds too personal to support any larger conclusions, but it tells me that to end a war you have to return to the same personal relationships you would have had with people before it. You do make peace. Nothing is constant in history.

8

Letting Go and Owning Up

A man travels many miles to consult the wisest guru in the land. When he arrives, he asks the great man: "O wise guru, what is the secret of a happy life?"

"Good judgment," says the guru.

"But, O wise guru," says the man, "how do I achieve good judgment?"

"Bad judgment," says the guru.

As we follow the trail of self-justification through the territories of family, memory, therapy, law, prejudice, conflict, and war, two fundamental lessons from dissonance theory emerge: First, the ability to reduce dissonance helps us in countless ways, preserving our beliefs, confidence, decisions, self-esteem, and well-being. Second, this ability can get us into big trouble. People will pursue self-destructive courses of action to protect the wisdom of their initial decisions. They will treat those they have hurt even more harshly, because they convince themselves that their victims deserve it. They will cling to outdated and sometimes harmful procedures in their work. They will support torturers and tyrants who are on the right side — that is, theirs. People who are insecure in their religious beliefs may feel the impulse to silence and harass those who disagree with them, because the mere existence of those naysayers arouses the painful dissonance of doubt.

But there is another side to dissonance: the pain that people feel when they *cannot* allow self-justification to erase the memory of the harms they caused, the mistakes they made, the decisions that backfired. That inability to let go can leave an indelible mark of regret and guilt, in extreme cases leading to despair, depression, or alcoholism. In soldiers, we call those symptoms PTSD. "How does a soldier justify his taking of life in the face of the powerful sanctions against this act that likely informed his upbringing?" asked psychologist Wayne Klug and his colleagues in their study of Iraq veterans. "Does his subsequent struggle with guilt, grief, and cognitive dissonance suggest a moral indictment of war?"[1]

Psychiatrist Jonathan Shay, who advises the military on post-traumatic stress disorders, observed that some veterans suffer continued "moral pain" over killings that they feel violated their ethical code, even if the killing was an inevitable part of war: "It occurs when you've done something in the moment that you were told by your superiors that you had to do, and believed, truthfully and honorably, that you had to do, but which nonetheless violated your own ethical commitments," he says. "There is a bright line between murder and legitimate killing that means everything to them . . . They hate it when they have killed somebody they didn't need to kill. It's a scar on the soul."[2]

The art of living with dissonance is as much about coping

with the scars on the soul as it is about avoiding them. Just as Odysseus had to steer his ship between Homer's mythical sea monsters Scylla and Charybdis — embodiments of rocky shoals and a whirlpool in the Strait of Messina, both perilous to sailors — so we must find a path between the Scylla of blind self-justification on one side and the Charybdis of merciless self-flagellation on the other. That middle course is more complex than letting ourselves off the hook right away with a quick defense — "What else could I have done?" or "It's the other guy's fault" or "I was just following orders" or "I wasn't wrong on the main point; just on a few details" or "Can we put this behind us and get back to business?" This tactic won't cut it, not with others and not with ourselves. It is important to stay on the hook for a while, to suffer some anguish, confusion, and discomfort on the road to understanding what went wrong. Only then can we gain an appreciation of what we have to do to make it right.

That process was certainly hard for Linda Ross, the psychotherapist who had practiced recovered-memory therapy until she realized how misguided she had been; for Grace, whose false recovered memories tore her family apart for years; for Thomas Vanes, the district attorney who learned that a man he had convicted of rape and who had spent twenty years in prison was innocent; for Vivian Gornick, who belatedly acknowledged her part in her history of failed relationships; for the couples and political leaders who eventually manage to break free of the spirals of rage and retaliation. And it is surely hardest of all for those whose professional mistakes cost the lives of friends and coworkers.

N. Wayne Hale Jr. was the launch integration manager at NASA in 2003, when seven astronauts died in the explosion of

the space shuttle *Columbia*. In a public e-mail to the members of
the space-shuttle program, Hale took full responsibility for the
disaster:

> I had the opportunity and the information and I failed
> to make use of it. I don't know what an inquest or a court
> of law would say, but I stand condemned in the court of
> my own conscience to be guilty of not preventing the *Co-
> lumbia* disaster. We could discuss the particulars: inatten-
> tion, incompetence, distraction, lack of conviction, lack
> of understanding, a lack of backbone, laziness. The bot-
> tom line is that I failed to understand what I was being
> told; I failed to stand up and be counted. Therefore look
> no further; I am guilty of allowing *Columbia* to crash.[3]

These courageous individuals take us straight into the heart
of dissonance and its innermost irony: the mind wants to protect
itself from the pain of dissonance with the balm of self-justification, but the soul wants to confess. To reduce dissonance, most of
us put an enormous amount of mental and physical energy into
protecting ourselves and propping up our self-esteem when it
sags under the realization that we have been foolish, gullible, mistaken, corrupt, or otherwise human. And yet, much of the time,
all this investment of energy is surprisingly unnecessary. Linda
Ross is still a psychotherapist — a better one. Thomas Vanes is still
a practicing attorney, perhaps a more thoughtful one. Grace got
her parents back. William Broyles found peace. N. Wayne Hale
was promoted to manager of NASA's space-shuttle program at
the Johnson Space Center, a position he held until his retirement.

The need to reduce dissonance is a universal mental mechanism, but, as these stories illustrate, that doesn't mean we are doomed to be controlled by it. Human beings may not be eager to change, but we have the ability to change, and the fact that many of our self-protective delusions and blind spots are built into the way the brain works is no justification for not trying. Is the brain designed to defend our beliefs and convictions? Fine — the brain also wants us to stock up on sugar, but most of us learn to eat vegetables. Is the brain designed to make us flare in anger when we think we are being attacked? Fine — but most of us learn to count to ten and find alternatives to beating the other guy with a cudgel. An appreciation of how dissonance works, in ourselves and others, gives us some ways to override our wiring. And protects us from those who can't. Or won't.

Mistakes Were Made — by Them

> These two gentlemen did not deserve what happened, and we are accountable. I am accountable.
> — *Starbucks CEO Kevin Johnson, after two African American men were arrested while waiting for a friend at a Starbucks*

Imagine, for a moment, how you would feel if your partner, your grown child, or your parent said: "I want to take responsibility for that mistake I made; we have been quarreling about it all this time, and now I realize that you were right, and I was wrong." Or if your employer started a meeting by saying, "I want to hear every possible objection to this proposal before we go ahead with it — every mistake we might be making." Or if you heard a district attorney at a press conference say, "I made a horrendous mistake. I failed to reopen a case in which new evidence showed that I and my office sent an innocent man to prison. I will apologize and this office will make amends, but that's not enough. I will also reassess our procedures to reduce the likelihood of ever convicting an innocent person again."

How would you feel about these people? Would you lose respect for them? Chances are that if they are friends or relatives, you will feel relieved and delighted. "My God, Harry actually admitted he made a mistake! What a generous guy!" You're not alone. In one study, 556 people were asked to read a scenario in which a pedestrian was injured by a speeding bicyclist. They were asked to imagine that they were the injured party and would be negotiating a settlement with the bicyclist. In one version of the scenario, the pedestrian received no apology; in another, a sympathetic apology ("I am so sorry that you were hurt. I really hope that you feel better soon"); and in a third, a responsibility-accepting apology ("I am so sorry that you were hurt. The accident was all my fault. I was going too fast and not watching where I was going until it was too late"). Participants who got the responsibility-accepting apology evaluated the bicyclist more positively, were more likely to forgive the bicyclist, and were more likely to accept a reasonable settlement.[4]

If the person who admits mistakes or harm is a business or political leader, you will probably feel reassured that you are in the capable hands of someone big enough to do the right thing, which is to learn from the wrong thing. The last American president to tell the country he had made a mistake that had disastrous consequences was John F. Kennedy in 1961. He had trusted the claims and faulty intelligence reports of his top military advisers, who assured him that once Americans invaded Cuba at the Bay of Pigs, the Cuban people would rise up in joy and overthrow Castro. The invasion was a disaster, but Kennedy learned from it. He reorganized his intelligence system and decided that he would no longer accept uncritically the claims of his military advisers, a change that helped him steer the country successfully through the subsequent Cuban missile crisis. After the Bay of Pigs fiasco, Kennedy said: "This administration intends to be candid about its errors. For as a wise man once said, 'An error does not become a mistake until you refuse to correct it.' ... Without debate, without criticism, no administration and no country can succeed — and no republic can survive." The final responsibility for the failure of the Bay of Pigs invasion was, he said, "mine, and mine alone." As a result of that admission, Kennedy's popularity soared.

That story sure feels like ancient history, doesn't it? Imagine a president apologizing and gaining respect and admiration for doing so! The legal scholar Cass Sunstein found in his studies that for many people today, "apologies are for losers." They can backfire, because if you don't like the person apologizing, you take his or her words as evidence of weakness or incompetence.[5] Moreover, given a national climate in which offenders *must* admit wrongdoing, express remorse, and promise repentance or they'll lose their jobs, their roles in a show, or their academic careers, apologies themselves have become polarized and politicized. When are they important, and when not? And for what behavior should they be offered? Many people are as dismayed by forced apologies for actions they personally find unobjectionable as they are by failures to apologize for behavior they find reprehensible.

Whatever the error, sin, or mistake, apologies fail when listeners know that the speaker has to say *something* to reassure the public but the statement feels formulaic and obligatory (which it often is, having been generated by a press agent or someone in human resources). That's a sure sign that the speaker doesn't really believe the apology is warranted and is self-justifying. Most of us are not impressed when leaders offer the form of an apology without its essence, saying essentially, "I didn't do anything wrong myself, but it happened on my watch, so, well, I guess I'll take responsibility."[6] We are not persuaded when CEOs apologize with vague hand-waving, as in Apple's non-apology apology over the performance of iPhone batteries. "We've been hearing feedback from our customers about the way we handle performance for iPhones with older batteries and how we have communicated that process," the company said. "We know that some of you feel Apple has let you down. We apologize." As Lisa Leopold, who has studied the language of apologies, wondered, what were they apologizing for, the poor-performing batteries, their bad communication process, or the feelings of their customers?[7]

When there is incontrovertible evidence of wrongdoing, the public longs to hear authorities own up, without weaseling or blowing smoke, followed by the next part: "And I will do my best to ensure that it will not happen again." Daniel Yankelovich, the highly regarded survey researcher, reported that although polls find that the public has an abiding mistrust of the nation's major institutions, right below that cynicism is a "genuine hunger"

for honesty and integrity. "People want organizations to operate transparently," he says, "to show a human face to the outside
world, to live up to their own professed standards of behavior,
and to demonstrate a commitment to the larger society."[8]

An example of that hunger underlies the movement in the
health-care system to encourage doctors and hospitals to admit
and correct their mistakes. Traditionally, most doctors have been
adamant in their refusal to admit mistakes in diagnosis, procedure, or treatment on the self-justifying grounds that doing so
would encourage malpractice suits. They are wrong. Studies of
hospitals across the country have found that patients are actually less likely to sue when doctors admit and apologize for mistakes and when changes are implemented so that future patients
will not be harmed in the same way. "Being assured that it won't
happen again is very important to patients, more so than many
caregivers seem to appreciate," says Lucian Leape, a physician
and professor of health policy at the Harvard School of Public
Health. "It gives meaning to patients' suffering."[9]

Doctors' second self-justification for not disclosing mistakes
is that doing so would puncture their aura of infallibility and
omniscience, which, they maintain, is essential to their patients'
compliance and confidence in them. They are wrong about this
too. The image of infallibility that many physicians try to culti-
vate often backfires, coming across as arrogance and even heart-
lessness. "Why can't they just tell me the truth and apologize?"
patients and their families lament. The physician Atul Gawande
has written eloquently of "the problem of hubris" that afflicts
many doctors, their inability to admit they can't cure everything,
to talk straight to patients, to accept their limitations.[10] In fact,

when competent physicians come clean about their mistakes, they are still seen as competent but also as human beings capable of error. Richard A. Friedman beautifully summarized the difficulties and benefits of owning up. "Like every doctor," he began, "I've made plenty of mistakes along the way." In one case, he failed to anticipate a potentially dangerous drug interaction, and his patient ended up in the intensive care unit and almost died. "Needless to say, I was distraught about what had happened," he said. "I wasn't sure what went wrong, but I felt that it was my fault, so I apologized to the patient and her family. They were shaken and angry, and they quite naturally blamed me and the hospital . . . but in the end they decided this was an unfortunate but 'honest' medical error and took no legal action." The disclosure of fallibility humanizes doctors and builds trust, Friedman concluded. "In the end, most patients will forgive their doctor for an error of the head, but rarely for one of the heart."[11]

Recipients of an honest admission of error are not the only beneficiaries. When we ourselves are forced to face our own mistakes and take responsibility for them, the result can be an exhilarating, liberating experience. Management consultant Bob Kardon told us about the time he led a seminar at the National Council of Nonprofit Associations' conference. The seminar was entitled, simply, "Mistakes," and twenty leaders of the statewide associations attended. Kardon told them that the only ground rule for the session was that each participant had to tell about a mistake he had made as a leader and not try to clean it up by talking about how he had corrected it — or dodged responsibility for it. He did not allow them to justify what they did. "In other words," he told them, "stay with the mistake":

As we went around the circle the magnitude of the mistakes burgeoned. By the time we reached the halfway point these executives were copping to major errors, like failing to get a grant request in on time and costing their organization hundreds of thousands of dollars in lost revenue. Participants would often get uncomfortable hanging out there with the mistake, and try to tell a redeeming anecdote about a success or recovery from the mistake. I enforced the ground rules and cut off the face-saving attempt. A half hour into the session laughter filled the room, that nearly hysterical laughter of release of a great burden. It got so raucous that attendees at other seminars came to our session to see what the commotion was all about.

Kardon's exercise illuminates just how difficult it is to say, "Boy, did I mess up," without the protective postscript of self-justification — to say "I dropped a routine fly ball with the bases loaded" rather than "I dropped the ball because the sun was in my eyes" or "because a bird flew by" or "because it was windy" or "because a fan called me a jerk." A friend returning from a day in traffic school told us that as participants went around the room, reporting the violations that had brought them there, a miraculous coincidence had occurred: Not one of them had broken the law! They all had justifications for speeding, ignoring a stop sign, running a red light, or making an illegal U-turn. He became so dismayed (and amused) by the litany of flimsy excuses that, when his turn came, he was embarrassed to give in to the same impulse. He said, "I didn't stop at a stop sign. I was entirely wrong and I got caught." There was a moment's silence, and then the room erupted in cheers for his candor.

There are plenty of good reasons for admitting mistakes, starting with the simple fact that you will probably be found out anyway — by your family, your company, your colleagues, your enemies, your biographer. But there are more positive reasons for owning up. Other people will like you more. Someone else may be able to pick up your fumble and run with it; your error might inspire someone else's solution. Children will realize that everyone screws up on occasion and that even adults have to say "I'm sorry." And if you can admit a mistake when it is the size of an acorn, it will be easier to repair than if you wait until it becomes the size of a tree, with deep, wide-ranging roots.

At work, institutions can be designed to reward admissions of mistakes as part of the organizational culture rather than making it uncomfortable or professionally risky for people to come forward. This design, naturally, must come from the top. Organizational consultants Warren Bennis and Burt Nanus offer a story about the legendary Tom Watson Sr., IBM's founder and its guiding inspiration for over forty years. "A promising junior executive of IBM was involved in a risky venture for the company and managed to lose over $10 million in the gamble," they wrote. "It was a disaster. When Watson called the nervous executive into his office, the young man blurted out, 'I guess you want my resignation?' Watson said, 'You can't be serious. We've just spent $10 million educating you!'"[12]

• • •

Dissonance theory demonstrates why we can't wait around for people to turn into Tom Watson or have moral conversions, personality transplants, sudden changes of heart, or new insights that will cause them to sit up straight, admit error, and do the right thing. Most human beings and institutions are going to do everything in their power to reduce dissonance in ways that are favorable to them, that allow them to justify their mistakes and maintain business as usual. They will not be grateful for the evidence that their methods of interrogation have put innocent people in prison for life. They are not going to thank us for pointing out to them why their study of some new drug, into whose development they have poured millions, is fatally flawed. And no matter how deftly or gently we do it, even the people who love us dearly are not going to be amused when we correct their fondest self-serving memory . . . with the facts.

Because most of us are not automatically self-correcting and because our blind spots keep us from knowing when we need to be, external procedures must be in place to correct the errors that human beings will inevitably make and reduce the chances of future ones. In hospitals across the country, indeed around the world, the simple requirement for doctors and nurses to observe prescribed checklists of steps that must be followed in surgery, emergency-room procedures, and aftercare has reduced normal human error and mortality rates.[13] In the legal domain, we have seen that mandatory electronic recording of all forensic interviews is one obvious and relatively inexpensive corrective to the confirmation bias; any bias or coercion that creeps in can be assessed later by independent observers. This is the impulse for the movement to equip police officers and their vehicles with cameras, which can help resolve disputes when officers are accused of using excessive force. However, even having cameras on every car, lamppost, cell phone, and police officer will not fully fix the problem. Because "believing is seeing," people can watch the same videos of the same events — such as the dozens of police killings of African Americans that spurred the Black Lives Matter movement — and come away with entirely different views of what they saw and who was to blame. The chokehold death of Eric Garner in Staten Island, who died gasping, "I can't breathe"; the shooting of Michael Brown in Ferguson, Missouri; the shooting of Tamir Rice, a twelve-year-old boy brandishing a toy pistol in a Cleveland park; the shooting of Philando Castile in his car, the aftermath of which was live-streamed by his girlfriend — many viewed these episodes as clear examples of unprovoked and excessive police brutality; others saw them as acceptable police behavior.

It is not only potential police bias we need to worry about; it is also prosecutorial bias. Unlike physicians, who can be sued for malpractice if they amputate the wrong arm, prosecutors generally have immunity from civil suits and are subject to almost no judicial review. Most of their decisions occur outside of public scrutiny, because fully 95 percent of the cases that the police hand over to a prosecutor's office never reach a jury. But power without accountability is a recipe for disaster in any arena, and in the criminal justice system, that combination permits individuals and even entire departments to do anything for a win, with self-justification to smooth the way.[14] (This is why the Center for Prosecutor Integrity, mentioned in chapter 5, is an important step in the right direction.) When district attorneys actively seek to release an inmate found to be innocent (as opposed to grudgingly accepting a court order to do so), it is usually because, like Robert Morgenthau, who reopened the Central Park Jogger case, they were not the original prosecutors and therefore have no need for self-justification. That is why independent commissions must often be empowered to investigate charges of corruption in a department or determine whether to reopen a case. Their members must have no conflicts of interest, no decisions to justify, no cronies to protect, and no dissonance to reduce.

Few organizations, however, welcome outside supervision and correction. If those in power prefer to maintain their blind spots at all costs, then impartial review boards must improve their vision — against their will, if it comes to that. Scientific and medical journals, aware of the taint on research when conflicts of interest are involved and having been deceived by a few investigators who faked their data, are instituting stronger measures to reduce the chances of publishing biased, corrupt, or fraudulent research. Many scientists are calling for greater transparency in the review process, the same solution that reformers of the criminal justice system are seeking. The ultimate correction for the tunnel vision that afflicts all of us mortals is more light.

Mistakes Were Made — by Me

It is considered unhealthy in America to remember mistakes, neurotic to think about them, psychotic to dwell upon them.

　— Lillian Hellman, playwright

Our national pastime of baseball differs from the society that spawned it in one crucial way: It keeps track of its mistakes. The box score of every baseball game, from the Little League to the Major Leagues, consists of runs, hits, and errors. Errors are not desirable, but all fans and players understand that they are unavoidable. Errors are inherent in baseball, as they are in medicine, business, science, law, love, and life. But before we can deal with them, we must first acknowledge that we have made them.

If letting go of self-justification and admitting mistakes is so beneficial to the mind and to relationships, why aren't more of us doing it? If we are so grateful to others when they do it, why don't we do it more often? Most of the time we don't do it because, as we have seen, we aren't even aware that we need to. Self-justification purrs along automatically, just beneath consciousness, protecting us from the dissonant realization that we did anything wrong. "Mistake? What mistake? I didn't make a mistake . . . The tree jumped in front of my car . . . And what do I have to be sorry about, anyway? Not my fault."

So what exactly are we supposed to do in our daily lives? Call an external review board of cousins and in-laws to adjudicate every family quarrel? Video all parents' interrogations of their teenagers? In our private relationships, we are on our own, and that calls for some self-awareness. Once we understand how and when we need to reduce dissonance, we can become more vigilant about the process and often nip it in the bud, catching ourselves before we slide too far down the pyramid. By looking at our actions critically and dispassionately, as if we were observing someone else, we stand a chance of breaking out of the cycle of action, followed by self-justification, followed by more committed action. We can learn to put a little space between what we feel and how we respond, insert a moment of reflection, and think about whether we really want to buy that canoe in January, really want to send good money after bad, really want to hold on to an opinion that is unfettered by facts.

Social scientists are finding that once people are aware of their biases, know how they work, and pay mindful attention to them — in effect, once they bring them into consciousness and say, "*There* you are, you little bastard" — they have greater power to control them. Consider the bias we discussed in the first chapter, naive realism: the bias to believe that we see things clearly and therefore have no bias. This bias is the central impediment to negotiations between any two individuals or groups in conflict who see things entirely differently. In a study with Jewish Israelis and Palestinian Israelis, making them aware of naive realism and how it operates was enough to lead even the most hawkish participants to *see the bias in themselves* and become more open to seeing the other side's point of view.[15] We aren't naive; we do realize it will take more than this modest intervention to resolve the Middle East conflict. But the point is that people are educable about their biases and about dissonance too.

In 1985, Israeli prime minister Shimon Peres was thrown into dissonance by an action taken by his ally and friend Ronald Reagan. Peres was angry because Reagan had accepted an invitation to pay a state visit to the Kolmeshohe Cemetery at Bitburg, Germany, to symbolize the two nations' postwar reconciliation. Because forty-nine Nazi Waffen-SS officers were buried there, the announcement of the proposed visit enraged Holocaust survivors and many others. Reagan, however, did not back down from his decision to visit the cemetery. When reporters asked Peres what he thought of Reagan's action, Peres neither condemned Reagan personally nor minimized the seriousness of the visit to Bitburg. Instead, Peres took a third course. "When a friend makes a mistake," he said, "the friend remains a friend, and the mistake remains a mistake."[16]

Consider for a moment the benefits of being able to separate dissonant thoughts as clearly as Peres did: People can remain passionately committed to their nation, religion, political party, and family while disagreeing with actions or policies they find inappropriate, misguided, or immoral. Friendships are preserved, not terminated in a huff; mistakes are not dismissed as unimportant but properly criticized and their perpetrators held responsible, even if the perpetrators are friends. In a 2017 YouTube video, Sarah Silverman spoke directly about the dissonance she felt over the sexual misbehavior of Louis CK, her dear friend of twenty-five years: "I need to address the elephant masturbating in the room," she began. The Me Too movement was long overdue, she said, and we were going to learn bad things about people we liked — and people we loved. "I love Louis, but Louis did these things. Both of those statements are true. So I keep asking myself, Can you love someone who did bad things? . . . I am at once very angry for the women he wronged and the culture that enabled it, and also sad, because he's my friend." It was vital, she concluded, to hold people accountable for their actions. And it was also vital to support and help the friends we loved.

Peres's third course can also help us navigate the eternal dilemma of how to respond to information that a beloved or admired artist is, or was, a son of a bitch, a racist, an anti-Semite, a homophobe, a pedophile, a sexual harasser, or a contemptible individual in some other way in private life. The 2019 documentary *Leaving Neverland* described Michael Jackson's long-running sexually abusive relationships with two boys, beginning when they were ages seven and ten. The film devastated and divided Jackson's legion of fans. One side reduced dissonance by denying the allegations, vilifying the credible men who told their stories, maintaining Jackson's innocence, buying ads, loudly criticizing the HBO film, and, in some cases, threatening those involved in its production; a lawsuit for "sullying Jackson's memory" was filed against the men who alleged they had been abused.[17] The other side reduced dissonance by swearing they would never again listen to Jackson's music or by trying to erase his legacy from pop culture; *The Simpsons* pulled an episode that had a character voiced by Jackson, and Louis Vuitton removed some Jackson-inspired pieces from its collection.

Amanda Petrusich, a music critic and passionate admirer of Jackson, wrote: "It is admittedly difficult, while watching 'Leaving Neverland,' to hold in mind two contradictory but equally imperative ideas: that victims should be believed, and that the accused are innocent until proved guilty. The first is wildly crucial if we wish to protect the disenfranchised from egregious abuses of power. The second remains the crux of the American criminal-justice system. Can these two ideas coexist?"[18]

Well, yes, they damned well better coexist, but an understanding of dissonance shows why that coexistence is often so very difficult. Petrusich continued: "Right now it feels as if they have to, which means that we are sometimes required to make personal choices about how we accept or dismiss the informa-

tion made available to us." Biographer Margo Jefferson did just
that. In an updated introduction to her book *On Michael Jackson,* she wrote, "We've long seen how charming and generous
he could be. Now we've also seen how calculating, selfish, and
gripped by demons he was. We can't erase or unknow that. We
can only accept it, acknowledge what it stirs in us — despair,
grief, anger, compassion — and try to turn it into wisdom."[19]

When the dissonance is caused by something we ourselves
did, it is even more vital to keep Peres's third way in mind: Articulate the cognitions and keep them separate. "When I, a decent,
smart person, make a mistake, I remain a decent, smart person
and the mistake remains a mistake. Now, how do I remedy what
I did?" By identifying the two dissonant cognitions that are causing distress, we can often find a way to resolve them construc-
tively or, when we can't, learn to live with them until we have
more information. When we hear about a sensational allegation
in the news, especially one in which sex is involved, we can resist the emotional impulse to hurl ourselves off that pyramid in
outraged support of the accused or the accuser. Instead of slot-
ting the story into an ideological framework — "Children never
lie"; "Believe survivors, even if they remember nothing"; "All
fraternity men are rapists" — we can do something harder and
more radical: wait for the evidence. If we don't and instead take
sides impulsively, it will be difficult to accept that evidence later
if it suggests that we were wrong, as happened in the McMartin preschool case (where, the nation later learned, children had
been pressured to report increasingly preposterous allegations of
abuse) or the Duke lacrosse case (where, the nation later learned,
a stripper's allegations of rape against a group of players were

false, and the district attorney was disbarred for prosecutorial misconduct). We can try to balance sympathy and skepticism. And then we can learn to hold our conclusions lightly, lightly enough so that we can let them go if justice demands that we do.

Becoming aware that we are in a state of dissonance can also help us make sharper, smarter, conscious choices instead of letting automatic, self-protective mechanisms resolve our discomfort in our favor. Suppose your unpleasant, aggressive coworker has just made an innovative suggestion at a group meeting. You could say to yourself, "An ignorant jerk like her could not possibly have a good idea about anything," and shoot her suggestion down in flames because you dislike the woman so much (and, you admit it, you feel competitive with her for your manager's approval). Or you could give yourself some breathing room and ask yourself: "Could the idea be a smart one? How would I feel about it if it came from my ally on this project?" If it is a good idea, you might support your coworker's proposal even if you continue to dislike her as a person. You keep the message separate from the messenger. In this way, we might learn how to change our minds before our brains freeze our thoughts into consistent patterns.

• • •

Mindful awareness of how dissonance operates is therefore the first step toward controlling its effects. But two psychological impediments remain. One is the belief that mistakes are evidence of incompetence and stupidity; the other is the belief that our personality traits, including self-esteem, are embedded and unchangeable. People who hold both of these ideas are often afraid to admit error because they take it as evidence that they are blithering idiots; they cannot separate the mistake from their identity and self-esteem. Although most Americans know they are supposed to say "We learn from our mistakes," deep down they don't believe it for a minute. They think that making mistakes means they are stupid. That belief is precisely what keeps them from learning from their mistakes.

About a fourth of the entire American adult population has been taken in by one scam or another, some of them silly, some serious: "You've won a brand-new Mercedes, and we'll deliver it to you as soon as you send us the tax on that amount first"; "We're offering you gold coins you can buy at a tenth of their market value"; "This miracle bed will cure all your ailments, from headaches to arthritis"; "Your nephew or grandchild is in dire medical straits in a foreign port and needs your money." Americans of all ages lose millions of dollars to telemarketing frauds, but old people are hit hardest, losing several times as much money in any given scam as young people.

Con artists know all about dissonance and self-justification. They know that when people who think of themselves as smart and capable are faced with the evidence that they spent thousands of dollars on a magazine-subscription scam (yes, those still exist) or were lured into a romance with a fraudulent but seductive online Romeo (or Juliet), few will reduce dissonance by deciding they aren't smart and capable. Instead, many will justify spending that money by spending even more money to recoup their sunk costs — their losses. This way of resolving dissonance protects their self-esteem but virtually guarantees their further victimization: "If only I subscribe to *more* magazines, I'll win the big prize," they say; or "I know it's unlikely that we fell in love by e-mail, but I'm sending money to help bring him over because this is the real thing"; or "Those nice, thoughtful people who made me the investment offer would never cheat me, and besides, they advertise on Christian radio." Some older people are vulnerable to reducing dissonance in this direction because many of them are already worried that they are losing it — their competence as well as their money. And they don't want to give their grown children grounds for taking control of their lives.

Understanding how dissonance operates helps us rethink our own muddles, and it's also a useful skill for helping friends and relatives get out of theirs. Too often, out of the best of intentions, we do the very thing guaranteed to make matters worse: We hector, lecture, bully, plead, or threaten. Anthony Pratkanis, a social psychologist who investigated how scammers prey on old people, collected heartbreaking stories of family members pleading with relatives who had been defrauded: "Can't you see the guy is a thief and the offer is a scam? You're being ripped off!" "Ironically, this natural tendency to lecture may be one of the worst things a family member or friend can do," Pratkanis says. "A lecture just makes the victim feel more defensive and pushes him or her further into the clutches of the fraud criminal." Anyone who understands dissonance knows why. Shouting "What were you *thinking?*" will backfire because it means "Boy, are you *stupid.*" Such accusations cause already embarrassed victims to withdraw further into themselves and clam up, refusing to tell anyone what they are doing. And what they are doing is investing more money or buying more magazines, because now they really have an incentive to get the family savings back, show they are not stupid or senile, and prove that what they were thinking was perfectly sensible.[20]

Therefore, says Pratkanis, before a victim of a scam will inch back from the precipice, he or she needs to feel respected and supported. Helpful relatives and friends can encourage the person to talk about his or her values and how those values influenced what happened while they listen uncritically. Instead of irritably asking "How could you possibly have listened to that creep?" you say, "Tell me what appealed to you about the guy that made you trust him." Con artists take advantage of people's best qualities — their kindness, politeness, and desire to honor their commitments, reciprocate a gift, or help a friend. Praising the victim for having these worthy values, says Pratkanis, even if they got the person into hot water in this particular situation, will offset feelings of insecurity and incompetence.

So embedded is the link between mistakes and stupidity in American culture that it can be shocking to learn that not all cultures share it. In the 1970s, psychologists Harold Stevenson and James Stigler became interested in the math gap in performance between Asian and American schoolchildren: by the fifth grade, the lowest-scoring Japanese classroom was outperforming the highest-scoring American classroom. To find out why, Stevenson and Stigler spent the next decade comparing elementary classrooms in the United States, China, and Japan. Their epiphany occurred as they watched a Japanese boy struggle with the assignment of drawing cubes in three dimensions on the blackboard. The boy kept at it for forty-five minutes, making repeated mistakes, as Stevenson and Stigler became increasingly anxious and embarrassed for him. Yet the boy himself was utterly unself-conscious, and the American observers wondered why they felt worse than he did. "Our culture exacts a great cost psychologically for making a mistake," Stigler recalled, "whereas in Japan, it doesn't seem to be that way. In Japan, mistakes, error, confusion [are] all just a natural part of the learning process."[21] (The boy eventually mastered the problem, to the cheers of his classmates.) The researchers also found that American parents, teachers, and children were far more likely than their Japanese and Chinese counterparts to believe that mathematical ability is innate; if you have it, you don't have to work hard, and if you don't have it, there's no point in trying. In contrast, most Asians regard math success like achievement in any other domain; it's a matter of persistence and plain hard work. Of course you will make mistakes as you go along; that's how you learn and improve.

Making mistakes is central to the education of budding scientists and artists of all kinds; they must have the freedom to experiment, try this idea, flop, try another idea, take a risk, be willing to get the wrong answer. One classic example, once taught to American schoolchildren and still on many inspirational websites in various versions, is Thomas Edison's reply to his assistant (or a reporter), who asked Edison about his ten thousand experimental failures in his effort to create the first incandescent light bulb. "I have not failed," he told the assistant (or reporter). "I successfully discovered ten thousand elements that don't work." Most American children, however, are denied the freedom to noodle around, experiment, and be wrong in ten ways, let alone ten thousand. The focus on constant testing, which grew out of the reasonable desire to measure and standardize children's accomplishments, has intensified their fear of failure. It is certainly important for children to learn to succeed, but it is just as important for them to learn not to fear failure. When children or adults fear failure, they fear risk. They can't afford to be wrong.

Research by psychologist Carol Dweck suggests one reason for the cultural differences that Stevenson and Stigler observed: American children typically believe that making mistakes reflects poorly on their inherent abilities. In Dweck's experiments, some children were praised for their efforts in mastering a new challenge; others were praised for their intelligence and ability ("You're a natural math whiz, Johnny"). Many of the children who were praised for their efforts even when they didn't get it right eventually performed better and liked what they were learning more than children who were praised for their natural abilities did. They were also more likely to regard mistakes and criticism as useful information that would help them improve. In contrast, children praised for their natural ability were more likely to care more about how competent they looked than about what they were actually learning.[22] They became defensive if they did not do well or if they made mistakes, and this reaction set them up for a self-defeating cycle: If they didn't do well, then, to resolve the ensuing dissonance ("I'm smart and yet I screwed up"), they simply lost interest in what they were learning ("I could do it if I wanted to, but I don't want to").

It is a lesson for all ages: the importance of seeing mistakes not as personal failings to be denied or justified but as inevitable aspects of life that help us improve our work, make better decisions, grow, and grow up.

• • •

Understanding how the mind yearns for consonance and rejects information that questions our beliefs, decisions, or preferences not only teaches us to be open to the possibility of error
but also helps us let go of the need to be right. Confidence is a
fine and useful quality; none of us would want a physician who
was forever wallowing in uncertainty and couldn't decide how
to treat our illness, but we do want one who is open-minded
and willing to learn. Nor would most of us wish to live without
passions or convictions, which give our lives meaning and color,
energy and hope. But an unbending need to be right inevitably
produces self-righteousness. When confidence and convictions
are unleavened by humility, by an acceptance of fallibility, people can easily cross the line from healthy self-assurance to arrogance. In this book, we have met many who crossed that line:
the psychiatrists who are certain that they can tell if a recovered memory is valid; the physicians and judges who are certain that they are above conflicts of interest; the police officers
who are certain that they can tell if a suspect is lying; the prosecutors who are certain that they convicted the guilty party; the
husbands and wives who are certain that their interpretation
of events is the right one; the nations that are certain that their
version of history is the only one.

   All of us will have hard decisions to make at times in our
lives; not all of them will be right, and not all of them will be
wise. Some are complicated, with consequences we could never
have foreseen. If we can resist the temptation to justify our actions in a rigid, overconfident way, we can leave the door open to
empathy and an appreciation of life's complexity, including the
possibility that what was right for us might not have been right

for others. "I know what hard decisions look like," says a woman we will call Janine.

> When I decided to leave my husband of twenty years, that decision was right for one of my daughters — who said, "What took you so long?" — but a disaster for the other; she was angry at me for years. I worked hard in my mind and brain to resolve that conflict and to justify what I did. I blamed my daughter for not accepting it and understanding my reasons. By the end of my mental gymnastics I had turned myself into Mother Teresa and my daughter into a selfish, ungrateful brat. But over time, I couldn't keep it up. I missed her. I remembered her sweetness and understanding, and realized she wasn't a brat but a child who had been devastated by the divorce. And so finally I sat down with her. I told her that although I am still convinced that the divorce was the right decision for me, I understood now how much it had hurt her. I told her I was ready to listen. "Mom," she said, "let's go to Central Park for a picnic and talk, the way we did when I was a kid." And we did, and that was the beginning of our reconciliation. Nowadays, when I feel passionate that I am 100 percent right about a decision that others question, I look at it again; that's all.

Janine did not have to admit that she made a mistake; she didn't make a mistake in terms of her own life. But she did have to let go of her need to insist that her decision was the right one for her children. And she needed to have compassion for the daughter who was hurt by her action.

Act 2: The Arduous Journey to Self-Compassion

There are no second acts in American lives.
— *F. Scott Fitzgerald*

One afternoon Elliot got into a lively discussion with his friend David Swanger, a distinguished poet, about F. Scott Fitzgerald's famous aphorism.

"It means we don't get second chances," said Swanger. "We don't recover from early failure. That's why every time a politician or athlete or other public figure has a comeback, some commentator uses that person's success to disprove the quote."

"Don't Americans go to the theater?" Elliot said. "There are *three* acts in a traditional play. His quote is not about second chances — it's much more interesting than that. Besides, Gatsby himself is the best example in American literature of a man reinventing himself. You don't think F. Scott Fitzgerald knew about comebacks?"

"Well, that's the common meaning," said Swanger.

"But in any classic play, act two is where the action is," said Elliot. "In life as in a play, you can't leap from act one to act three. We skip act two at our peril, for that's when we go through the turmoil of confronting our demons — the selfishness, immorality, murderous thoughts, disastrous choices — so that when we enter act three, we have learned something. Fitzgerald was telling us that Americans are inclined to bypass act two; they don't want to go through the pain that self-discovery requires."

When Elliot started teaching, in 1960, he used Fitzgerald's observation to make a point that the two of us now regard as the centerpiece of our views about living with dissonance. Act 1 is the setup: the problem, the conflict the hero faces. Act 2 is the struggle, in which the hero wrestles with betrayals, losses, or dangers. Act 3 is the redemption, the resolution, in which the hero either emerges victorious or goes down in defeat.[23] In his lectures, Elliot used *Death of a Salesman,* the quintessential American play, to make Fitzgerald's point that Americans would just as soon skip the part about the struggle. Willy Loman's older brother, Ben, Biff and Happy's impressively rich uncle, represents the American dream.

WILLY [*to his sons*]: Boys! Boys! Listen to this. This is your Uncle Ben, a great man! Tell my boys how you did it, Ben!

BEN: Why, boys, when I was seventeen I walked into the jungle and when I was twenty-one I walked out. And by God I was rich!

WILLY [*to the boys*]: You see what I been talking about? The greatest things can happen!

"What the hell happened in the jungle?" Elliot would ask his students. "That's where the story is! That's act two! How did Ben do it? How did he solve his problems? *How* did he get rich? Did he help people? Did he kill people? Did he lie, steal, cheat? What did he learn, and was it a lesson he can now offer his nephews?"

When the two of us were writing the first edition of this

book, we disagreed about whether to discuss the other side of dissonance, the suffering it creates in people who can neither justify nor forgive themselves for the harms they have caused or the bad decisions they have made. Elliot was opposed to our saying much about self-forgiveness because of his concern that people would miss the point of act two. "I don't want people short-circuiting the process," he said. "It's not enough to say, 'Hey, I did a bad thing and I won't do it again. It's important for me to forgive myself.' Yes, it is important, but the goal is not to use self-compassion as a Band-Aid to cover up the wound rather than take active steps toward its healing. People can go to confession, religiously or publicly, and admit they did a bad thing and they are sorry, but it won't make a dime's worth of difference if they don't *get* what that bad thing was and *get* that they are not going to do it again."

There is, in short, a big difference between superficial self-compassion and earned self-compassion. This distinction is especially important nowadays, because in recent years there has been a growing movement in positive psychology emphasizing the emotional, cognitive, and even motivational benefits of self-compassion. Who could be critical of this glowing concept? Yet it is more complex than it first appears, and it is easy to oversimplify it into a buzzword.

Psychologists Laura King and Joshua Hicks argue that maturity depends on the adult's capacity to confront lost goals, or lost possible selves, and acknowledge regrets and sorrows over roads not taken or dreams unfulfilled. "Lost possible selves," they write, "represent the person's memory of a self they would have pursued 'if only'" — if only my child did not have Down syn-

drome, if only I had been able to have children, if only my partner hadn't left me after twenty years. Reflecting on these lost
expectations poses costs to happiness — in our terms, it generates
painful dissonance — but, King and Hicks add, "that work, the
articulation of what might have been, may have benefits in terms
of the complexity of a person's sensibility and, perhaps, the very
meaning of happiness itself. That there is value in loss is more
than a platitude. Although it may be a peculiarly American in-
stinct to search for the positive in any negative event, we argue
that the active, self-reflective struggle to see the silver lining is
a key ingredient of maturity."[24] Exactly: maturity means an *active, self-reflective struggle* to accept the dissonance we feel about
hopes we did not realize, opportunities we let slide by, mistakes
we made, challenges we could not meet, all of which changed
our lives in ways we could not anticipate.

And to do this, we must apply the same compassion to ourselves that we would extend to others. King and Hicks found
that the people in their sample who had the lowest psychological well-being perceived their earlier selves as "foolish," "misguided," or "stupid," and could see no benefits or gains to the
losses of their dreams. Those with the highest well-being, however, had been able to take what the researchers describe as "an
unusually brutal perspective on a former self ": "Should I say
I was an idiot?" said one woman. "I had no idea what the life
was that I was dreaming about." Yet she can now look back at
that lost self with compassion, a self who can be excused for her
naïveté. The happiest, most mature adults were those who could
embrace the losses in their lives and transform them into sources
of deep gratitude — not with platitudes or Pollyanna glosses, say

the researchers, but by discovering the genuinely positive aspects of their multifaceted lives.

How, though, can we forgive ourselves for actions we consider unforgivable? Causing an innocent person's death is the most extreme mistake people can make; as Jonathan Shay said, it leaves a "scar on the soul." Consider the stories of two men who sought to treat that scar in different ways.

When Reggie Shaw was nineteen, he was texting as he drove his SUV to work and, distracted, he crossed the yellow divider and clipped a car coming the other way. The car spun out of control and crashed, killing the other driver and his passenger. For two years, until his trial for negligent homicide, Reggie Shaw denied any culpability. Then he listened to the testimony of scientists describing the psychology of distraction — how the brain responds to the demand for split attention, how reinforcing texting is, how it impairs accurate perception of danger; in short, how technology can, in one expert's words, cause "neurological hijacking." The more that Reggie learned about the science and confronted the other evidence that indisputably convicted him, says Matt Richtel, who wrote a book on the case, "the more he transformed into a zealot against the use of phones behind the wheel." Reggie was sentenced to only a brief stay in jail and community service, but the sentence he imposed on himself was much more severe. Ever since, Reggie has been telling his story to anyone who will listen, including high-school kids, athletes, policymakers, and legislators. "I'm here for one reason," he begins his talks. "That's for you guys to look at me ... and say: 'I don't want to be that guy.'" The prosecutor who brought the case against Reggie told Richtel: "I have never seen anybody try to redeem themselves as much as Reggie Shaw. Period. End of story." The judge added, "He's done more to effect change than anyone I've ever seen."[25]

Some people don't want the scar on the soul to heal over. They see it as a reminder of what they did, a protest against apathy or forgetting. One of the most touching examples we encountered was an essay by Eric Fair, who had been a contract interrogator in Iraq in 2004 at Abu Ghraib. Ten years later he was teaching a college writing class. "The course's title, Writing War, kept me from straying too far from the memories that have haunted me over the last decade," he wrote. "I tortured. Abu Ghraib dominates every minute of every day for me." And then he showed his students the iconic photos of the tortured detainees. "As I looked at their blank faces, I realized I could let myself feel a powerful sense of relief," he said. "Abu Ghraib will fade. My transgressions will be forgotten. But only if I allow it."

Eric Fair has no intention of allowing it. After confessing to the U.S. Army's Criminal Investigation Command, he went on to speak about his actions to any audience that invited him. "I've said everything there is to say. It's not hard to pretend the best thing to do is put it all behind me."[26]

> I stood before the class that day tempted to let apathy soften the painful truths of history. I no longer had to assume the role of the former interrogator at Abu Ghraib. I was a professor at Lehigh University. I could grade papers and say smart things in class. My son could ride the bus to school and talk to his friends about what his father does for a living. I was someone to be proud of.

But I'm not. I was an interrogator at Abu Ghraib. I tortured.

Fair chooses not to reduce dissonance over his actions at Abu Ghraib by putting it all behind him and justifying what he did as something he had to do, under orders, as part of his job. Instead, he has chosen to bear witness to history, to the ugliest parts of the human psyche. He wants to remind his students that "this country isn't always something to be proud of." He doesn't want to forget. He doesn't want to forgive himself. That's his moral choice.

Yet it seems to us that Eric Fair is in the middle of his personal act 2, not engaging in repeated or purposeless self-flagellation but rather *wrestling with his demons*. By revealing what he did at Abu Ghraib to one class of students after another, he is working his way to a resolution, one in which he need not forget what he did but in which the memory does not dominate "every minute of every day" for him. His own words illuminate how he might get there: "I tortured," he says. He does not say "I am a torturer." In this way he separates his behavior from his identity, and that ability is what ultimately allows people to live with behavior they now condemn. His son may not be proud of what his father did in Iraq, but he can certainly be proud of his father's courageous honesty and determination to make amends — starting with teaching his family, his students, and his fellow citizens the lesson he suffered to learn.

All of us can carry this understanding into our private lives: something we did can be separated from who we are and who we want to be. Our past selves need not be a blueprint for our future selves. The road to redemption starts with the understanding that who we are *includes* what we have done but also *transcends* it, and the vehicle for transcending it is self-compassion.

Getting to true self-compassion is a process; it does not happen overnight. It does not mean forgetting the harm or error, as in "Ah, well, I'm basically a good, kind person, so I'll treat myself gently and move on." No; you might be a good, kind person but you are one who committed a grievously harmful act. That's part of you now, of who you are. But it need not be all of you. It need not define you — unless you keep justifying that act mindlessly.

In the previous chapter, we discussed the nation's response to the Senate Intelligence Committee's indictment of the CIA for brutality and deceit in its program of "enhanced interrogation" techniques. Instead of demanding the hard work of reforming the CIA, many members of Congress and political commentators were quick to say, in effect, that was act 1. It's what we did after 9/11. It's in our past. What's done is done. The mistakes can't be remedied now; we're in act 3. No one illustrated this attitude better than Fox news commentator Andrea Tantaros. "The United States of America is awesome," she said. "*We* are awesome. But we've had this discussion. We've closed the book on it. The reason they [the Senate Intelligence Committee members] want the discussion is not to show how awesome we are. It's to show us how we're not awesome. They apologized for something."[27] (She dismisses torture as *something?*) Now, if Andrea Tantaros understood that her country needs to spend a little time in act 2, that would be awesome.

• • •

Dissonance may be hardwired, but how we think about mistakes is not. After the disastrous bloodbath of Pickett's Charge at the Battle of Gettysburg, in which more than half of his 12,500 men were slaughtered by Union soldiers, the Confederate general Robert E. Lee said, "All this has been my fault. I asked more of my men than should have been asked of them."[28] Lee was a great general who made a tragic miscalculation, but that mistake did not make him an incompetent military leader. If Robert E. Lee could take responsibility for an action that cost thousands of lives, surely all those people in traffic school can admit they ran a red light.

There are a few Robert E. Lees in our modern military. Retired lieutenant general Daniel Bolger, who commanded troops in Iraq and Afghanistan between 2005 and 2013, published a mea culpa. "By the enemy's hand, abetted by my ignorance, my arrogance, and the inexorable fortunes of war," he wrote, "I have lost eighty men and women under my charge, with more than three times that number wounded. Those deaths are, as Robert E. Lee said at Gettysburg, all my fault":

> As generals, we did not know our enemy — never pinned him down, never focused our efforts, and got all too good at making new opponents faster than we could handle the old ones . . . we backed into not one but two long, indecisive counterinsurgent struggles to which our forces were ill-suited. Time after time, as I and my fellow generals saw that our strategies weren't working, we failed to reconsider our basic assumptions. We failed to

question our flawed understanding of our foe or ourselves … In the end, all the courage and skill in the
world could not overcome ignorance and arrogance. As
a general, I got it wrong. And I did so in company with
my peers.[29]

General Bolger's willingness to face up to the immense
magnitude of the disaster that the United States unleashed in
the Middle East is inspiring. But just as Vivian Gornick's life
might have been vastly happier had she not waited until age sixty-five to see her role in her failed relationships, we are entitled
to wish that General Bolger and his fellow officers had spoken
up sooner. The worsening of disasters might be averted if our
political and military leaders could change direction when the
course they are on is headed over a cliff. General Bolger leaves
the solution to fighting the "war on terror" to the next generation of military leaders, just as, he says, the American military
"took an uncompromising look at itself" after the failures in
Vietnam. "Good ideas and bad, lessons learned, re-learned, and
unlearned — all deserve thorough scrutiny and discussion," he
writes. Yes, they do. But it's late; possibly too late. We can't keep
fighting the past war, let alone the lost war. What is needed is a
deep understanding not only of what went wrong then but also
of what is going wrong right now, the better to prepare for *what
could go wrong* with current decisions. We need an Eisenhower
strategy.

   In June 1944, Dwight Eisenhower, supreme commander of
the Allied forces in Europe, had to make a crucial military deci-

sion. He knew the invasion of Normandy would be costly under the best of circumstances, and the circumstances were far from ideal. If the invasion failed, thousands of troops would die in the effort, and the humiliation of defeat would demoralize the Allies and hearten the Axis powers. Nonetheless, Eisenhower was prepared to assume full responsibility for the possibly catastrophic consequences of his decision to go forward. He wrote out a short speech he planned to release if the invasion went wrong. It read, in its entirety:

> Our landings in the Cherbourg-Havre area have failed to gain a satisfactory foothold and the troops have been withdrawn. My decision to attack at this time and place was based upon the best information available. The troops, the Air [Force] and the Navy did all that bravery and devotion to duty could do. If any blame or fault attaches to the attempt, it is mine alone.[30]

After writing this note, Eisenhower made one small but crucial change. He crossed out the end of the first sentence — "the troops have been withdrawn" — and replaced that passive construction with "and I have withdrawn the troops." The eloquence of that *I* echoes down the decades.

In the final analysis, a nation's character, and an individual's integrity, do not depend on being error free. It depends on what we do after making the error. The poet Stephen Mitchell, in his poetic rendering of Chinese philosopher Lao Tzu's *Tao Te Ching,* writes:

A great nation is like a great man:
When he makes a mistake, he realizes it.
Having realized it, he admits it.
Having admitted it, he corrects it.
He considers those who point out his faults
as his most benevolent teachers.

9

Dissonance, Democracy, and the Demagogue

The man who lies to himself and listens to his own lie comes to such a pass that he cannot distinguish the truth within him, or around him, and so loses all respect for himself and for others. And having no respect he ceases to love, and in order to occupy and distract himself without love he gives way to passions and coarse pleasures, and sinks to bestiality in his vices, all from continual lying to other men and to himself.

— *Fyodor Dostoyevsky,* The Brothers Karamazov

December 24, 2019

Dear Reader: This chapter can best be considered a work in progress. Originally we decided to conclude this edition of our book with a chapter about Donald Trump, because no application of dissonance theory is more important than understanding how his presidency has further widened what seems to be an unbridgeable chasm between political parties, friends, and members of the same family.

Unfortunately, it takes months for a finished manuscript to become a book, and that is why, by the time you read this, you will know a good deal more about Trump and his fate than we do now at the end of 2019 (which may seem like a lifetime ago to you).

We have no idea what will become of Trump, given the volatility of his personality and his presidency. A bitterly divided House of Representatives passed two articles of impeachment against Trump: abusing the power of his office by pressuring a foreign power, Ukraine, to meddle in our elections by digging up dirt on his political rival Joe Biden, and obstruction of Congress, for refusing to cooperate with the congressional hearings and permit key aides to testify. In the House Intelligence Committee's hearings, ambassadors, National Security Council officials, and members of the State Department and the Foreign Service testified that Trump had done exactly what he was accused of doing and that everyone in his inner circle knew it. As cognitive dissonance theory would predict, Republicans — who unanimously opposed the resolution on impeachment and did not dispute the facts presented — called the hearings a sham and a hoax.

Some of our friends and colleagues thought we were crazy to try to write anything about the political scene today, since whatever we wrote would be outdated in a day, a week, a month, a year. As one friend put it, "He could be removed from office, reelected, defeated, start a war with Iran, cause a civil war here — who can say?"

We can't, of course. But as social scientists, we do have a great deal to say about how the case study of Donald Trump sheds light on a larger issue — the Trump phenomenon. In 2016, sixty-three million Americans voted for him, some with full-throated enthusiasm and others with mistrust and doubt, all of them hoping that he would be the president to meet their political, economic, and emo-

tional needs — that he would keep factories humming and make deals with foreign countries to benefit the economy and that an administration of professionals would rein in his personal excesses. The overwhelming majority of Trump's supporters, having made this initial commitment to him with their votes, remain loyal to him in spite of the dissonance generated by his increasingly outrageous and erratic behavior, his litany of lies, and his inflammatory, divisive rhetoric. In this chapter, therefore, we will focus not only on Trump but also on his unwavering followers; we will show you how their escalating self-justifications can erode the soul of a nation and its fundamental institutions. As dissonance theory would also predict, only a minority of Trump's supporters have changed their minds about him, and we believe it is crucial to understand how and why they did, given the personal, professional, and psychological costs that many of them paid.

• • •

Let's start with a story from history that may serve as a parable for our times. It is the tale of a basically decent man who, with the best of intentions, agreed to endorse the political excesses of a powerful leader in order to achieve the humane ends he sought. How could that decision go wrong? But it did. One step at a time.

Most people know about Pope Pius XII and his collaboration with the Nazis during World War II. Fewer know about the connection between his predecessor Pope Pius XI, who was elected pope in 1922, and Benito Mussolini, who became the Italian prime minister that same year. Pius XI and Mussolini had little in common other than a pervasive Italian Catholic anti-

Semitism, and they met only once in the seventeen years between the pope's election and his death. Mussolini was no friend
of the Catholic Church; as a young man he was called *mangiaprete* (priest-eater), and later his Fascist squads regularly attacked
priests and terrorized members of Catholic Action, a network of
religious youth clubs. Ever since Italy's creation as a nation-state,
in 1861, the country had emphasized liberal and secular values,
and the pope feared that Mussolini would continue his assaults
on the Church. Pius XI was not a Fascist, however; in 1926, he
instituted a ban on Catholic participation in the right-wing,
proto-Fascist Action Française, led by France's foremost anti-
Semite.

In his Pulitzer Prize–winning history *The Pope and Mussolini,* David Kertzer details the tactics by which Mussolini
triumphed over the Church.[1] Mussolini, knowing that the Vatican's approval would play a major role in legitimizing his violent Fascist regime, began systematically wooing the pope to his
side as soon as he became prime minister. In his first speech to
his new parliament, he pledged to build a Catholic state befit-
ting a Catholic nation and asked for God's help — something no
leader had done since the nation's founding. Pius XI was somewhat reassured, but still apprehensive. "If he could be sure Mussolini would work to restore Church influence in Italy," writes
Kertzer, "he was not inclined to hold his anticlerical past against
him. . . . Never under any illusion that Mussolini personally embraced Catholic values or cared for anything other than his own
aggrandizement, the pope would be willing to consider a pragmatic deal if he could be convinced that Mussolini would deliver
on his promises."[2] The first step.

Mussolini set about proving that he was a good Catholic. He ordered his cabinet to kneel in prayer at the altar of the Unknown Soldier in Rome. He had his children and his wife (who despised the Catholic Church) baptized. He paid for the restoration of churches that had been damaged in the Great War. He required crucifixes to be placed in courts, hospitals, and classrooms. He, the former *mangiaprete*, made it a crime to insult a priest. He decreed that all elementary schools must teach the Catholic religion. He accommodated the pope's war on "heresy," banning Protestant books, a biography of Cesare Borgia, and journals that the Vatican found offensive. Mussolini assured Pius XI that he would do nothing to the Jews that the Church had not already done. And so the pope, pleased to achieve the religious ends that the Vatican wanted, stifled any concerns about Mussolini's Fascist means of asserting power. The second step, and Pius XI was hooked.

All the while, Mussolini continued to unofficially endorse violence committed by his supporters against Catholic Action. These youth groups were dear to Pius XI's heart, for he saw them, as Kertzer puts it, as "ground troops for re-Christianizing Italian society." The pope was outraged by these attacks, but Mussolini "proved adept at using the violence to his benefit, convincing the pope that he was the only man in Italy who could keep the rowdies under control."[3] The culprits were rarely even arrested, let alone punished.

In 1929, the Vatican and the Italian state signed official accords. Pius XI was happy because, among other gifts to the Church, the accords specified that the Catholic religion was "the only religion of the state." Mussolini was happy because the accords silenced any Catholics who believed or hoped that the pope opposed the Fascist regime. Mussolini now lacked any significant opposition, and his craving for adulation ballooned. Before long, he was demanding that schoolchildren pray to him, Il Duce, and offer their lives to him rather than to God — an act of true heresy from the Church's perspective, but the pope did not protest. Priests and bishops were summoned from all over the country to celebrate Mussolini's agricultural policy, and, fearing to offend him, off they went, marching through Rome to lay wreaths not at Catholic shrines but Fascist monuments. The priests were required to cheer Mussolini's entrance at a public ceremony and pray for blessings upon him. In 1935, Mussolini even induced Pius XI to bless his genocidal invasion of Abyssinia (now Ethiopia) by calling it a "holy war"; one hundred thousand Italian soldiers were sent into battle as a distraction from Italy's economic woes.

Throughout the 1930s, the pope struggled to justify the benefits the Church was getting from Mussolini despite his growing alarm about the rise of Nazism and its virulent anti-Semitism. The pope was not concerned about a "Jewish threat" in Italy; he was far more worried about the Nazi threat in Europe. To keep the pope from speaking out against anti-Semitism, Mussolini persuaded him that the Italian version was different from the Nazi version, and besides, Mussolini said, he wouldn't treat Jews any more savagely than the Church itself had. Accordingly, the pope began distinguishing between "good Fascism," which recognized the Church's rights, and "bad Fascism," which did not. In 1937, Kertzer writes, Mussolini bragged to the German foreign minister about how easy it had been to manipulate the Church. Just allow religious education in the schools, he advised. True, there had been a little trouble with the Catholic Action groups, but he had quickly brought the Vatican in line. All it took was doing "small favors for the higher clergy," such as giving them a few tax concessions and free railway tickets.

By the mid-1930s, the pope could no longer live with the precarious balance between means and ends that he had created for himself. As Peter Eisner, author of *The Pope's Last Crusade,* put it, "The pope realized that today it was the Jews, but then it would be the Catholics and finally the world. He could see in the day's news that the Nazis would stop at nothing less than world domination."[4] Pius XI enlisted the support of an American Jesuit who had written about racism, asking him to draft a papal encyclical that would publicly condemn Hitler, Mussolini, and their goal to exterminate all Jews. On his deathbed, the pope prayed to live a few more days so that he could deliver a speech with the, dare we say, truly Christian message that eventually "all peoples, all the nations, all the races, all joined together and all of the same blood in the common link of the great human family," would be united by faith. The pope also planned to condemn "the prohibition of marriages between Aryans and non-Aryans." It was too late. He died the next day, without giving the speech.

His successor Eugenio Pacelli, who would shortly become Pope Pius XII, ordered the pope's secretary to gather all notes pertaining to the speech, and he instructed the Vatican's printer, who had the text ready for distribution, to destroy every copy. This the printer did, assuring Pacelli that "not a comma" remained.* The new pope, his nuncio reported to Mussolini, spoke "with much sympathy for Fascism and with sincere admiration for the Duce." Almost immediately, Pius XII lifted the ban on Catholics joining the Action Française.

• • •

In telling the story of Pope Pius XI, we do not intend to imply that Donald Trump is Benito Mussolini. But the two men do have one important thing in common: They exhibit all the classic characteristics of a demagogue, starting with the qualities of grandiosity and an unquenchable need for praise. Both wooed their natural adversaries with blandishments, rewards, and a few sops to achieve their goals. Each offered himself as the only leader who could solve the nation's problems — "Nobody knows the system better than me," Trump said in accepting his party's nomination, "which is why I alone can fix it." Demagogues thrive and flourish on the reasoning and self-justifications of those willing to push aside their moral objections in exchange for political advantage. And, above all, demagogues exploit public prejudices and ignorance, fomenting anger and hatred at the expense of reasoned argument.

Americans are well aware of the deadliest demagogues of the twentieth century — Hitler, Stalin, Mussolini — but we have long endured our share of homegrown examples. The historian Robert Dallek argues that America's demagogues — notably

---

* Twenty years later, after Pope Pius XII died, Pope John XXIII released excerpts of the speech, excising passages critical of the Fascist regime. The full text was not released until 2006.

Louisiana governor Huey Long in the 1930s, Wisconsin senator Joseph McCarthy in the 1950s, and Alabama governor George Wallace in the 1960s — "can be seen as predecessors of Trump's ascent to political power."[5] But none of these men attained the presidency of the United States.

Demagogues, by definition, need adoring crowds, and they create them by using the timeless method of arousing fear. At his nomination speech at the Republican National Convention, Trump offered a litany of things to fear: violent crime, lawless migrants, "men, women, and children viciously mowed down" by terrorists, the rising crime rate (a lie; nationwide, crime has been declining for decades), and the "damage and destruction" plaguing our cities. "I have a message for all of you," he said. "The crime and violence that today afflicts our nation will soon come to an end. Beginning on January twentieth, 2017, safety will be restored." How? He would do it — alone. "I have visited the laid-off factory workers and the communities crushed by our horrible and unfair trade deals," he said. "These are the forgotten men and women of our country. People who work hard but no longer have a voice. *I am your voice.*"

Demagogues typically thrive by sowing division among citizens and inciting scapegoating and violence, and no American president before Trump fomented us-versus-them thinking to such an extreme degree, much less tacitly endorsed violence by "us" against "them." He refers to the free press, the very bedrock of a democracy, as an "enemy of the people," which has led some of his supporters — apparently thinking it was funny — to wear T-shirts that say NOOSE. TREE. JOURNALIST. SOME ASSEMBLY REQUIRED. And then there's his constant inveighing against immigrants, those "thugs" and "animals" who are "invading" our country. "How do you stop these people? You can't," he said at rally in Florida in May of 2019. Someone in the crowd yelled, "Shoot them." The audience of thousands cheered. Trump smiled and joked, "Only in the Panhandle can you get away with that statement." Two months later, a white supremacist murdered twenty people in El Paso to halt the "Hispanic invasion of Texas."[6]

The election of a demagogue to the White House has arguably been the greatest internal threat to our democracy since the Civil War, and justifying Trump's behavior requires far more contortions than supporting George Bush's disastrous Iraq War did. Although no one can predict the outcome for this particular demagogue, history gives us a pretty good idea of what happens to a nation that falls under the sway of one: it doesn't turn out well. The rise of any demagogue never happens overnight, and it's never the result of one election. It occurs because of the slow shift in beliefs and values that follows every self-justifying decision that citizens make. One step at a time.

The Pyramid of Choice, Once Again

The guiding metaphor of this book has been the pyramid of choice: As soon as people make a decision, whether reasoned or impulsive, they will change their attitudes to conform to that choice and start minimizing or dismissing any information suggesting they chose the wrong option. Typically, in politics, people let their party identity make the decisions for them, which is why most voters feel little dissonance in supporting the candidate who heads their own party — "I'm a Republican [or Democrat]; that's who I am and who I vote for." But what happens when that candidate holds beliefs or behaves in ways that formerly would have been anathema to those voters?

Once upon a time, the Republicans were virulently anti-Communist and regarded the former Soviet Union as an "evil empire" (Ronald Reagan's term) and Russia as an ideological enemy of capitalism, and they would brook no "radical left" criticism of the FBI or CIA. How did so many Republicans come to tolerate an American president's cozy friendship with Vladimir Putin, and why do they fail to become enraged at the evidence of Russia's meddling with an American presidential election? How could they shift from endorsing the Cold War slogan "Better Dead Than Red" to proclaiming "Better Dead Than Democrat"? How did so many forget that it was Richard Nixon who signed the Clean Air Act and decide to support a party that wants to dismantle all environmental protections? How did so many sit silent as the demagogue roared?

By the time of the presidential election in 2016, American citizens had plenty of information about Donald Trump: his promise to control the alleged hordes of Mexican rapists and other criminals who were illegally flooding the country; his insults directed at ethnic minorities, disabled people, and women; his falsely claiming to have foot problems in order to avoid being drafted during the Vietnam War; his long history of refusing to pay his contractors for their work; his bankruptcies and refusal to release his tax returns (lying that he couldn't because he was being audited, which the IRS says is nonsense); his decades-long history of discrimination against African Americans who worked for him or tried to rent apartments in his buildings; the extramarital affairs that ended his marriages; the accusations of sexual misconduct by numerous women; and his vulgar comment to *Access Hollywood* host Billy Bush that "when you're a star, you can do anything [with women] — grab 'em by the pussy." Any *one* of these facts would once have doomed a candidate's chances, but Trump thrived, much to his own apparent surprise: "I could stand in the middle of Fifth Avenue and shoot somebody and wouldn't lose any voters, okay? It's, like, incredible." Not incredible, of course, to anyone who understands cognitive dissonance.

Most people who get caught in a lie, mistake, or hypocritical dance feel sharp dissonance and are motivated to squirm out of it with a flurry of self-justifications. But Trump has always been unfazed precisely because he feels no dissonance when caught. Feeling dissonance requires the ability to feel shame, guilt, empathy, and remorse, and he lacks that capacity. The only justifications he feels the need to make are claims that he can do anything he wants because he is "a very stable genius" and knows more than anyone about everything. He can never learn from his mistakes because he convinces himself that he never makes any — he cuts dissonance off at the pass, never allowing it into his brain. He is the quintessential con man, someone for whom lying is second nature. It's just what you do. If the chumps believe you, that's their problem.[7]

So imagine that there you are in 2016, a lifelong Republican or a Democrat who can't stand Hillary Clinton, confronted with a candidate unlike any who has ever achieved the highest office in the land. You are at the apex of the pyramid facing a decision.

Which way do you jump? Vote for Trump enthusiastically because, as one of his admirers said, "He speaks from the heart and
speaks his mind," and what he speaks is what you feel? Vote for
him because, as a lifelong Republican, you know he will follow
the Republican agenda, even if he does come with a few character flaws you would have loathed in a Democrat? Vote for him
because, although you dislike his vulgarity, sexual affairs, and
prejudices, he shares your stance on an issue you are passionate about, such as abortion, immigration, or Israel? Hold your
nose and vote for him, whatever your party loyalty, because at
least he's not "crooked Hillary," a term the Trump campaign has
been repeating in speeches and Facebook ads for months?* Vote
for him because, frankly, you feel angry and scared about all the
changes you see around you, including unfamiliar ethnic groups
gaining political ground and deteriorating conditions in your
hometown? Don't vote at all?

It's often hard to remember that throughout the 2016 primaries, the majority of Republican voters preferred any of the other
seventeen candidates who were hoping for the nomination. (In
the first round of primaries, for example, only about one-third
of voters supported Trump, and two-thirds favored the others.[8]) There was never a typical "Trump voter," though it later
seemed so to many observers. Trump voters diverged on a variety of issues, including taxes, entitlements, immigration, race,
gay marriage, gender equality, and other social issues, and many
had previously voted for Barack Obama. In spite of Republicans'

---

* One ad featured two handcuffs as the *o*'s in *crooked,* and of course Trump basked in the
repeated chants of "Lock her up" at his rallies.

doubts and diverse feelings about Trump *before* the election, however, their support for Trump *after* his election grew and rarely wavered. As of mid-2019, nearly 90 percent of Republicans approved of Trump's performance in office, even though 65 percent also said they considered his conduct "unpresidential."

How does a person justify support for an unpresidential president? Easy. If you were on the fence about him, you are likely to jump over it to his side now, because after all, you voted for him, and if you voted for him, you want your vote to be consistent with your feelings about him today. Yet your brother-in-law keeps telling you that his election was a disaster and what the hell were you thinking? Thanks to the cognitive biases that ensure that people see what they want to see and seek confirmation of what they already believe, you now, post-election, are motivated to focus on what you like about him and dismiss what you don't like. You tune in more regularly to Sean Hannity, who assures you that you did the right thing. Besides, you never did think much of your brother-in-law anyway. In this way, you minimize any discomfort you feel that you might have made the wrong decision.

Let's begin, then, with the supporters who originally felt considerable dissonance about voting for him but did it anyway — and then found themselves working hard to justify that decision. (And let's not forget the self-justifications of the 48 percent of eligible voters who didn't bother to cast a ballot, the ones who said, "My vote doesn't make a difference," or "I'm sick of voting for the lesser of two evils.") Demagogues come and go, but self-justifications are forever.

Stepping Off the Pyramid: "The Democrats Are Worse"

Shortly after Trump announced his presidential candidacy, a Never Trump movement, spearheaded by Mitt Romney and other prominent Republicans, tried vigorously to thwart his nomination. When it became clear they had failed and that Trump would be the party's nominee, some, notably the distinguished conservative columnist George Will, immediately stepped off the pyramid by officially leaving the Republican camp. "This is not my party," said Will, adding that he would henceforth be unaffiliated. "Make sure he loses," he advised his fellow conservatives.

Another establishment Republican, Ari Fleischer, the former press secretary to George W. Bush, went back and forth about what to do. At first he declared, "There's a lot about Donald Trump that I don't like, but I'll vote for Trump over Hillary any day." As the campaign wore on, he changed his mind, writing in a *Washington Post* op-ed two weeks before the election that Trump had "veered recklessly off track, attacking an American judge for his Mexican heritage, criticizing a war hero's family, questioning the legitimacy of the election and otherwise raising questions about his judgment."[9] Fleischer decided to leave his ballot blank.

The great majority of Republicans, even Trump's noisiest opponents, did not follow these examples of leaving the party or not voting. Almost all politicians, understandably, criticize their opponents in the primaries, sometimes harshly and sometimes with flair, but they eventually support the party's nominee. Most of Trump's Republican opponents, however, were unusually outspoken in their personal animus toward him and in their fear for their party and the country. Ted Cruz, the target of a particularly nasty barrage of insults from Trump (he had tweeted that Cruz's wife was ugly and, preposterously, that Cruz's father was involved in the assassination of JFK), said, "We need a commander in chief, not a Twitterer-in-chief. I don't know anyone who would be comfortable with someone who behaves this way having his finger on the button. I mean, we're liable to wake up one morning, and Donald, if he were president, would have nuked Denmark." He would never endorse Trump, he said, because "history isn't kind to the man who holds Mussolini's jacket."[10]

That was nothing compared to Lindsey Graham's comments on him prior to the election: "You know how you make America great again? Tell Donald Trump to go to hell." He added:

> He's a race-baiting, xenophobic, religious bigot. He doesn't represent my party. He doesn't represent the values that the men and women who wear the uniform are fighting for.... I don't think he has a clue about anything. He's just trying to get his numbers up and get the biggest reaction he can. He is helping the enemy of this nation. He is empowering radical Islam. And if he knew anything about the world at all, [he] would know that most Muslims reject [radical] ideology.[11]

My, how far they have fallen. By the midterm elections of 2018, Trump could not have had more sycophantic endorsers than Cruz and Graham. Cruz, evidently willing to not only hold Mussolini's jacket but also give its wearer a bear hug, did just that at a Trump rally in Texas. By 2019, Graham, the man who had called Trump a "kook," a "con man," a "complete idiot," and a "race-baiting, xenophobic, religious bigot," was denying that Trump was a racist or that his supporters' chant to "Send [Ilhan Omar] back" was racist, and besides, he said on *Fox and Friends,* those congresswomen are "Communists" who hate America. How did Cruz, Graham, and thousands of other Republican politicians and opinion leaders get to that point?

All presidents lie, fudge, and dissemble — or, at the very least, play fast and loose with the truth — and all are loath to admit mistakes and deceptions. John F. Kennedy did not write *Profiles in Courage,* a book for which he received much admiration and a Pulitzer Prize; his speechwriter Ted Sorensen wrote most of it. Richard Nixon told the nation, "I am not a crook," even though he was, and Bill Clinton said, "I did not have sex with that woman," even though he did. Ronald Reagan lied when he claimed that his administration had not secretly arranged an illegal sale of arms to Iran and used the money to fund the Contras in Nicaragua. Barack Obama, promoting his Affordable Care Act in 2013, told the country, "If you like your health care plan, you can keep it" — a statement that PolitiFact called its Lie of the Year. People who hear these lies are, thanks to dissonance reduction, inclined to minimize or justify those that *their* politicians make as being trivial or understandable or otherwise excusable.

Yet it became evident from the first day of Trump's presidency that the self-protecting falsehoods of his predecessors would pale in comparison to his own, starting with the inflated numbers of

people he claimed were at his inauguration. It is not coincidental that in 2016, the year that Donald Trump was elected, Oxford
Dictionaries anointed as its word of the year *post-truth*, which it
defined as "relating to or denoting circumstances in which objective facts are less influential in shaping public opinion than
appeals to emotion and personal belief" — another tactic of the
demagogue, and the Trump administration began implementing
it immediately. After Trump's inauguration in 2017, his adviser
Kellyanne Conway said on *Meet the Press* that press secretary
Sean Spicer didn't actually *lie* to the press that day; he had simply called upon "alternative facts," a statement that evoked widespread laughter. "I ain't cheating," one online commenter posted,
"I have an alternative girlfriend." "I had to double check that this
wasn't an SNL skit," said another.

But public amusement was followed by no small degree of
horror when it turned out she wasn't kidding.* The term *alternative facts* may be ludicrous but it is serious, especially when it is
used to justify an administration's denial of global warming, the
environmental harms of coal and pesticides, and other well-estab-
lished scientific findings. Alternative facts, even when soundly re-
futed by credible experts, persist because they are sticky; that is,
each repetition of a falsehood makes it more familiar and therefore more believable.[12] All demagogues know this.

---

* Kellyanne's husband, conservative George Conway, has consistently tweeted his disgust with Trump's behavior and "pathological" lies. Clearly this couple works hard to resolve any dissonance caused by their differing views. Kellyanne said her commitment to
Trump is a "feminist" act because no feminist would approve of her giving up her job on
account of her husband's beliefs. George said on a talk show that his marriage was no different from the countless others in Washington in which spouses disagree.

Though many critics have drawn a comparison between Trump's alternative facts and the propaganda technique that Hitler, in *Mein Kampf,* called "the big lie" (meaning a lie so huge that no one would believe that anyone "could have the impudence to distort the truth so infamously"), we think historian Zachary Jonathan Jacobson expressed an equally vital concern: "What we should fear today," he wrote, "is not the Big Lie but the profusion of little ones: an untallied daily cocktail of lies prescribed not to convince of some higher singularity but to confuse, to distract, to muddy, to flood. Today's falsehood strategy does not give us one idea to organize our thoughts, but thousands of conflicting lies to confuse them."[13]

For Republicans who felt any discomfort in voting for Trump, here was evidence that made their cognitive dissonance worse: the daily demonstrations that Trump lied as easily as he breathed and that he was utterly incapable of acknowledging that he was wrong. On July 16, 2018, standing next to Vladimir Putin at a press conference in Helsinki, Trump sided with Putin and disputed his own intelligence community's conclusion that Russia had meddled in the 2016 election. "I will say this, I don't see any reason why it would be [Russia]," said Trump. "President Putin was extremely strong and powerful in his denial today." U.S. intelligence officials and members of both parties, including Senator John McCain, Republican strategist Newt Gingrich, and some Fox News anchors, were outraged, and some called his behavior treasonous. The next day, at the White House, Trump tried to backtrack. He had full faith in America's intelligence agencies, he said. "I accept our intelligence community's conclusion that Russia's meddling in the 2016 election took place" (although he could not stop himself from adding, "It could be other people also; there's a lot of people out there"). He was unfazed by the fact that less than twenty-four hours previously, he had said the exact opposite. "I thought it would be obvious but I'd like to clarify just in case it wasn't. In a key sentence in my remarks, I said the word *would* instead of *wouldn't.*" See, a double negative, he joked; he meant to say "'I don't see any reason why it *wouldn't* be Russia.' . . . I think that probably clarifies things pretty good." No, it didn't, because the context in which Trump made his initial remarks was unambiguous; it could not possibly have been a slip of the tongue. Even his most ardent apologists shook their heads in dismay at this lame attempt at an explanation.

Trump lies about trivial matters just as often as significant ones, again because he simply cannot be wrong. On March 11, 2019, at a roundtable meeting of business executives, Trump referred to Apple's CEO Tim Cook as "Tim Apple." It was funny and it was unimportant, but Trump couldn't let it go. Within a few days, he was claiming he'd never even said it: "I quickly referred to Tim + Apple as Tim/Apple as an easy way to save time & words," Trump tweeted. Later he told a group of donors that he'd actually said "Tim Cook Apple" really fast, so that's why no one heard the *Cook* part.[14]

The *Washington Post* reported that by April 19, 2019, Trump had passed the ten-thousand-lie mark, and the newspaper proceeded to list every one of them; that number, inflated by Trump's fury over the impeachment investigation, was up to 15,413 by December 17 . . . and counting.[15] PolitiFact rated 69 percent of his statements as "mostly false or worse" and only 17 percent as "mostly true." Trump lies when he changes his mind, claiming he didn't change his mind. He makes things up; for example, by claiming that wind power doesn't work, that wind turbines are "killing all the eagles" and cause cancer, that he went to Ground Zero shortly after 9/11 and sent hundreds of men to help the rescue effort (he didn't). He repeats claims he knows are false (for example, that Barack Obama was not born in the United States and that Democratic congresswoman Ilhan Omar married her brother). He invents stories to inflame his base, as when he claimed that Democrats endorse infanticide by allowing doctors to "execute" newborns.* He denies his actions even when there is clear evidence on video that he did what he denies.[16] He doctored an official weather map with a Sharpie — a federal crime, by the way — to make it seem like Hurricane Dorian would hit Alabama, a clumsy attempt to cover his mistaken claim that it would reach that far west.

You are a Trump supporter. How do you reduce dissonance when faced with the evidence of your president's troubling, outrageous, and foolish falsehoods? You minimize their importance, because "all presidents lie." Or you find them funny. Or you deny the whole issue. Does he lie? Don't be absurd, said Stuart Varney, host of Fox Business Network. He has never lied to the American people. He merely "exaggerates and spins."[17]

Some supporters may have assuaged their dissonance prior to the election by assuring themselves that his erratic behavior would be controlled by competent White House staff and cabinet members. If so, that hope was dashed by the ensuing chaos

---

* "The baby is born. The mother meets with the doctor. They take care of the baby. They wrap the baby beautifully. And then the doctor and the mother determine whether or not they will execute the baby." Said at a rally in Green Bay, Wisconsin, April 28, 2019.

in the Trump administration. Within two years, the Brookings Institution reported, he had a senior staff turnover rate of 43 percent, and *Business Insider* began keeping a running tally of all the top-level people who had been fired or who "resigned."* "A high turnover rate is a signal of rot and decay within any organization," wrote Stephanie Denning in *Forbes,* hardly a left-wing mouthpiece.[18]

Moreover, while it is not unusual for politicians to disagree with a president in their own party or even say negative things about him, the vitriol directed against Trump is unique in American history—and it has come from his employees, friends, allies, and members of his inner circle, including his secretary of state, his secretary of defense, his national security adviser, and his chief of staff.[19] They have called him "a supreme sexist," "like an 11-year-old child," "morally unfit and untethered to truth," "less a person than a collection of terrible traits," "dumb as shit," someone who "sucks up and shits down" (this from former Fox News chief Roger Ailes), "not only crazy but stupid," and "an idiot."

Members of Trump's cabinet, staff, political party, or social world thus faced significant dissonance upon realizing that their leader undeniably had profound liabilities of personality, cognitive function, and competence.[20] The options for reducing that dissonance were clear: Stay or go. Support Trump, and get the payoff of the continued prestige, power, and possible national

---

* Anthony Scaramucci holds the record for the shortest tenure. Although he was ousted after only eleven days as White House communications director, he remained a loyal Trump supporter for two more years before announcing in an op-ed that he'd finally had enough.

importance of your high-level job; dissent, and pay the price of incurring Trump's wrath, being exiled, or losing your seat in Congress. Some, perhaps caught between conscience and constituents, retired from politics early. Some stayed but kept their critical faculties and integrity intact — even when, like Rex Tillerson, they thought Trump was a "fucking moron" — out of ethical and patriotic concerns. Hence the curious 2018 op-ed in the *New York Times* titled "I Am Part of the Resistance Inside the Trump Administration" in which the anonymous author tried to reassure the country that many Trump appointees had vowed to do what they could "to preserve our democratic institutions while thwarting Mr. Trump's more misguided impulses until he is out of office. The root of the problem is the president's amorality. Anyone who works with him knows he is not moored to any discernible first principles that guide his decision making."[21] In other words, the country needs to know that there are a few adults in the room and that *somebody* in the White House will put on the brakes.

Sliding Down the Pyramid: "He Has Eaten Your Soul"

By 2019, it was apparent that there were no brakes; anyone who dared to argue with Trump's decisions and whims, let alone criticized him, disagreed with him, called his attention to a mistake he'd made, or tried to control his "misguided impulses," was ousted for what Trump considered disloyalty. Seeing disagreement as disloyalty is another hallmark of demagogues, dictators, and strong-arm leaders. Indeed, after special counsel Robert

Mueller published his report outlining eleven episodes in which Trump or his administration had tried to have the special counsel fired or limit or interfere with the scope of his investigation, observers inside and outside the White House began comparing Trump to a Mafia Mob boss for whom loyalty is the ultimate value. For example, the report states that on June 17, 2017, Trump called Don McGahn, the White House counsel, and ordered him to turn the screws on the deputy attorney general, Rod Rosenstein. "Call Rod, tell Rod that [Robert] Mueller has conflicts and can't be the special counsel," Trump said. "Mueller has to go . . . Call me back when you do it." Underlings are expected to do what it takes to "fix" the boss's problems. That's what Michael Cohen did — and went to prison for. He paid Stormy Daniels $130,000 in hush money immediately before the election to keep their affair quiet (a violation of federal campaign finance laws) and perjured himself about the payment in an attempt to protect the president. Fixers are expected to obstruct justice, lie, pay off sexual partners who might make trouble, and break the law on their boss's orders.

"[Trump] conducts himself like a New Jersey mob boss who is unconcerned about asking the people around him to conduct unethical or legally challenging behaviour," said Kurt Bardella, former spokesperson and senior adviser for the House Oversight and Government Reform Committee. "Truth and accuracy just don't factor into his thought process at all. The demands for loyalty and fealty are like an organized crime network. Instead of the John Gotti family, it's the Trump family and his soldiers are the Republican members of Congress who protect him."[22] And what are the signature crimes of a Mafia soldier? Being a rat and flipping on your boss. Trump called President Nixon's whistleblower John Dean a "rat" for telling the truth about the Watergate crimes and coverups, but Trump had a special antipathy toward flippers. Having worked with Mafia figures throughout his business career, Trump said on Fox News, "I know all about flipping, for thirty, forty years I've been watching flippers. Everything's wonderful, and then they get ten years in jail and they flip on whoever the next highest one is, or as high as you can go." But flipping has long been a centerpiece of law enforcement procedures, because it is how prosecutors can get to the ultimate perpetrators: If you catch a hit man with a smoking gun, you want the murderer, but even more you want to know who ordered the hit. The method, therefore, is aimed at getting lower-level criminals to tell the truth about their bosses. Yet what is Trump's view of this fundamental strategy of law and order? Flipping is dishonorable, Trump said on Fox News, and so unfair it "almost ought to be outlawed."[23]

If flipping is disloyal, whistleblowing is treason, according to Trump. On August 12, 2019, an intelligence officer in the CIA filed an official complaint regarding the president's behavior, a complaint that set in motion the House impeachment inquiry.

> In the course of my official duties, I have received information from multiple U.S. Government officials that the President of the United States is using the power of his office to solicit interference from a foreign country in the 2020 U.S. election. This interference includes, among other things, pressuring a foreign country to investigate one of the President's main domestic political rivals. The

President's personal lawyer, Mr. Rudolph Giuliani, is a central figure in this effort. Attorney General Barr appears to be involved as well.

Trump accused this intelligence officer and others who supported the complaint of being spies and traitors. "I want to know who's the person who gave the whistleblower the information because that's close to a spy," Trump said. "You know what we used to do in the old days when we were smart? Right? With spies and treason, right? We used to handle them a little differently than we do now" — clearly implying execution. He can't actually behead or assassinate dissenters and "traitors" as Putin and Kim Jong-Un can, but he can send dog whistles to his henchmen, as with his tweet that if he were removed from office there would be a "Civil War like fracture in this Nation." He accused Representative Adam Schiff, chair of the House Intelligence Committee's Ukraine investigation, of treason for simply paraphrasing the transcript of that phone call and said Schiff should be arrested. And when Senator Mitt Romney tweeted, "By all appearance, the President's brazen and unprecedented appeal to China and to Ukraine to investigate Joe Biden is wrong and appalling," Trump tweeted that Romney was a "pompous ass" and called for his impeachment (despite the fact that senators cannot be impeached).

How did Trump's supporters initially respond to the CIA officer's complaint that the president had asked a foreign leader to do him a favor and get some dirt on his adversary Joe Biden? How did they then respond to the two weeks of confirming testimony from conscientious and well-informed witnesses during the House impeachment hearings? A few, like Romney, issued full-fledged condemnations; others hesitantly acknowledged that Trump's behavior was "troubling" or "not appropriate"; but most remained silent or mindlessly recited the White House's talking points about conspiracies, hoaxes, and witch-hunts. They had spent a lot of mental effort justifying all those earlier outrageous things Trump did, so what was one more? Lindsey Graham's first response was "to impeach any president over a phone call like this is insane." Yet Graham must know full well that the grounds for impeachment include soliciting foreign help in an election, obstruction of justice, witness intimidation, and violation of the statute guaranteeing the privacy and safety of whistleblowers.

It is conceivable that Graham and Ted Cruz privately hold their original views of Trump; most professional politicians learn to stifle their personal feelings for the sake of being re-elected and party loyalty. But we strongly suspect that because of the extraordinary gap between their pre-election feelings and their post-election capitulation, they have found a way to justify that transformation — if only to get a good night's sleep. Ari Fleischer, who originally favored Trump, then opposed Trump, then ended up not voting, is now an enthusiastic Trump supporter who frequently goes on Fox News and CNN to attack Trump's critics. How does he justify his change of heart? By saying, essentially, "The Democrats are worse." As he told Isaac Chotiner of the *New Yorker,* "When I add up Donald Trump, as offensive as he can be, as inappropriate as he can be, and combine it with his policy accomplishments, and I compare it to the Democrats and their statements and their policies and their rush to the far left, I will take Donald Trump any day."[24]

For supporters who cared only about a single issue — for example, appointing conservative judges to the Supreme Court, getting tax cuts, or backing Israel — Trump's delivery on that issue was all that mattered. A Republican donor on the board of the Republican Jewish Coalition, when asked about the divisive tensions Trump had exacerbated among American Jews who were critical of Benjamin Netanyahu's hardline policies, said, "My God, when I look at what he's done for Israel, I'm not going to take issue with anything he's said or done."[25] What about the rising number of anti-Semitic hate crimes in America that Trump's rhetoric has facilitated, comments such as his claim that there were "good people" among those at the Charlottesville rally who were chanting "Jews will not replace us"? No dissonance there: Trump can't be anti-Semitic; he has a Jewish son-in-law.

•   •   •

And so, within three years of Trump's election, most Republicans had slid to the bottom of a vast self-justifying pyramid. As Peter Wehner lamented, the Republican Party had become "Donald Trump's party, through and through."[26] Chief *Politico* correspondent Tim Alberta, author of *American Carnage,* detailed the capitulation of many Trump opponents who had once considered themselves die-hard Never Trumpers. Most had slid down the pyramid in the direction of "always Trump," having silenced any murmurings of conscience about his amorality, volatile moods, lies, and caprices, maybe by telling themselves, "He's not so bad after all. What the hell, he did put two conservatives on the Supreme Court, he's made dozens of appointments of conservatives to lower courts, he gave us an enormous tax cut, and he is gutting those damned environmental regulations that have stifled business. The mass shootings are terrible, of course, and we wish he'd tone down his rhetoric. But if he doesn't, well, he may be a demagogue, but he's our demagogue."

When the Supreme Court ruled that Trump and his administration would not be allowed to ask a question about citizenship on the 2020 census, Attorney General William Barr, in a public event in the Rose Garden, assured Trump that the whole issue was merely a "logistical" matter having to do with timing and applauded him for courageously agreeing to abide by the Court's decision, declaring, "Congratulations again, Mr. President." Congratulations for what? For accepting the Supreme Court's decision, which all presidents are obligated to do under the Constitution? Congratulations for not throwing yet another temper tantrum?

William Barr himself is an example of a man who slid down the pyramid without a bump, going from being a respected attorney to becoming the nation's attorney general to turning into Trump's sycophantic personal lawyer, traveling the world to induce foreign governments to accept Vladimir Putin's denial of responsibility for election interference. By the time members of the administration join Barr at the bottom of the pyramid, how likely are any of them to change their minds, question the president's lapses in judgment, and try to trek back up to the top? Consider Stephanie Denning's plaintive question in her col-

umn in *Forbes,* referring to the many Trump appointees who had
jumped ship (voluntarily or otherwise): "From a citizen's vantage
point, the comportment, crassness, and chaos synonymous with
the administration would surely warrant some sort of admission
of regret for joining it in the first place," she wrote. "Apparently
not. Why won't anyone admit they were wrong?"[27]

Why? Because by the time you are at the bottom of the pyramid, admitting you were wrong means acknowledging that you
sacrificed your better judgment to your immediate self-interest
or that you — smart, politically savvy, professionally experienced
you — failed to control, constrain, or even influence your thin-
skinned boss. It would mean that all your earlier justifications
were ... mistaken. Thanks to dissonance reduction, many of
Trump's loyalists, even those who were sacked, will not see themselves as sellouts or facilitators; they will convince themselves
that the Republican agenda — and that fat tax cut that put many
dollars in their pockets — are worth the small price of lavishing a
few compliments on him and turning a blind eye to his offenses.
Congratulations indeed.

For those at the bottom of the pyramid, the ultimate dissonance-reducing justification is, of course, that the ends justify the means. Political scientist Greg Weiner observed that
Trump's "most strenuous apologists have long swept all [com-
plaints against him] away with the breezy assurance that he
should be taken 'seriously, not literally,'" because he gives them
policies they want, and that serves the greater national good
or their religious agenda. Weiner identified an even more con-
voluted stroke of self-justification: that Trump could not have
achieved his policy agenda *without* the lying, vulgarity, and il-

licit behavior, so it all actually enhanced his effectiveness. This is what Weiner called the "post Trump, ergo propter Trump" fallacy: "It is a form of the 'post hoc ergo propter hoc' error in logic: 'after, therefore because of.' The classic illustration is the supposition that the rooster's crow causes the sunrise because the second event follows the first. In the version of the fallacy his defenders espouse, Mr. Trump violates customary standards of presidential behavior and then delivers desired policies, so the assumption is that the violations produced the policies."[28]

And what's wrong with the idea that "the ends justify the means"? Presidents of all parties, including Franklin D. Roosevelt, who had the Great Depression and World War II to deal with, have resorted to that default excuse. But when the "means" involve a trampling of the norms, rules, customs, and traditions of a democracy, Weiner argued, *that* is the legacy that will endure, not the momentary policies they justify. The more egregious and immoral the means, the more Trump's supporters must justify them as being essential to his ends — his policy agenda. This spurious reasoning infuriates Weiner. There is nothing, he said, that Trump has achieved for which his incivility and "unpresidential behavior" have been indispensable or even useful.

You may recall from chapter 1 how Jeb Stuart Magruder, an active participant in the Watergate scandal, described the steps of self-justification that led him to the bottom, sinking in a morass of corruption and crime in the Nixon administration. In May 2019, James Comey described the same descent for those in the Trump administration, articulating why it is so hard to go back up toward the apex of the pyramid. Comey, the former FBI director and author of *A Higher Loyalty: Truth, Lies, and*

Leadership, was dismissed by Donald Trump over his failure to state publicly that Trump was not under investigation for collusion with the Russians during the 2016 election. Comey began by wondering how it came to be that a "bright and accomplished lawyer" like Barr, as well as other prominent figures in Trump's orbit, could end up channeling the president's refrain of "No collusion" and accuse the FBI of "spying" on him. How, Comey asked, could Barr, testifying before the Senate Judiciary Committee, minimize Trump's attempt to fire Robert Mueller before he completed his report?[29] One step at a time.

First, you sit silently while he lies, makes false assertions, and creates "a web of alternative reality" that you and your associates do not dispute. You are pulled into "a silent circle of assent." You don't challenge his boast that he had the largest inauguration crowd in history, and you sympathize when he whines that he has been treated very unfairly by the press.

Second, you go along with his grandiose demand that you praise him publicly and swear your loyalty to him at cabinet meetings and other public venues. "You do what everyone else around the table does," wrote Comey, "you talk about how amazing the leader is and what an honor it is to be associated with him."

Third, you are silent as Trump attacks the values you cherish and the institutions you have vowed to protect. You are silent because what can you say? He's the president of the United States. You are bothered by his "outrageous conduct" but you stay because you feel you are needed to protect those values and institutions. You are too important to quit.

Comey concluded:

You can't say this out loud — maybe not even to your family — but in a time of emergency, with the nation led by a deeply unethical person, this will be your contribution, your personal sacrifice for America. You are smarter than Donald Trump, and you are playing a long game for your country, so you can pull it off where lesser leaders have failed and gotten fired by tweet.

Of course, to stay, you must be seen as on his team, so you make further compromises. You use his language, praise his leadership, tout his commitment to values.

And then you are lost. He has eaten your soul.

Landing at the Base: "We Have Seen His Heart"

Die-hard Trump loyalists, of course, were never on the fence; they voted for him for precisely the reasons that caused dissonance for so many others. For them, the confirmation bias is amplified by the megaphone of Fox News, which routinely transforms his vices into virtues and even provides prepackaged justifications for the party faithful to repeat frequently: "He's not a politician"; "He's rough around the edges"; "He has good intentions"; "He's not politically correct." And the classic self-justification — minimizing the evidence: "He has faults," said one woman at a rally, "but don't we all?"

Is he unpresidential? Being unpresidential isn't important, they say. In fact, it's the very reason we voted for him. A woman at a Women for Trump rally in Pennsylvania told the *Philadelphia Inquirer:* "He gets us. He's not a politician, and he's got a backbone. He's not afraid to say what he thinks. And what he says is what the rest of us are thinking."[30] At another rally in South Carolina, a sixty-four-year-old retired nurse said, "Everything he says is how I feel. I know this president. I've been to his inauguration, been to his other rallies. Everything he says I agree with. He's speaking for me. He may be a little rough around the edges, but he's not a politician."

Does he say some embarrassing things in his daily tweets taunting anyone who displeases him, from Rosie O'Donnell and Bette Midler to diplomats and heads of state? Sure, but that's nothing to be upset about. A woman at a Trump rally said, "Everybody's tweeting crazy things. Everybody is! Why point the finger at him?" (Um, because he is the president?)[31]

Does he throw a fit when he feels insulted or criticized? "He's not always the best at how he handles his emotions," said a thirty-two-year-old man. "He's a very emotional guy. Passionate. But I like his policies and I think he has good intentions."

Is he a racist? Of course not, said another young male supporter when asked about Trump's demonizing of four minority Democratic congresswomen. "He's just making a point and speaking his mind. That's important. There aren't enough people who say that nowadays. Everyone is politically correct. You can't get out what you want to say. I like that in a person."

And what about his erratic behavior, the administration turnover, the Mueller investigation, the impeachment hearings, and his own staff's comments about what an idiot he is? We noted that Trump supporters within the administration and party are fully aware of this uncomfortable information but tend to minimize it by saying, as Ari Fleischer did, that the Democrats are worse. But Trump's most loyal supporters wave away any potential whiff of dissonance with the simple claim that the unwelcome information is all fake news. The Mueller investigation into the corruption of the 2016 election? Just a ploy by Democrats to keep our guy from doing his job. Indeed, as dissonance theory predicts, *the stronger and more persuasive the criticism, the stronger and more entrenched the need to ignore it.* Trump's critics "throw shit at him every day, all day long," a sixty-nine-year-old real estate agent told a reporter, and then — unintentionally nailing the final step in reducing dissonance — she added: "It makes us want to support him more."[32]

• • •

When Trump promised his audience at the Republican National Convention that he would be the voice of the "forgotten men and women of our country," many heard the promise of keeping factories open and carrying out the conservative agenda, but a large minority heard something else: a man promising to allay their anxieties about real and imagined dangers and changes in the world. A sixty-nine-year-old retired respiratory therapist told a reporter that "he wants to protect this country, and he wants to keep it safe, and he wants to keep it free of invaders and the caravan and everything else that's going on. He understands why we're angry, and he wants to fix it."[33]

Political scientist Ashley Jardina, author of *White Identity Politics,* has studied that anger, and much of it, she found, stems from the misperception that white people are getting a disproportionately low share of the nation's resources. Many white voters support Trump, Jardina learned, because they are thinking

"Hey, Trump is there for my group. He's going to help white people. He's the president for white people."[34] Moreover, given the emotional power of their need to believe he is the "president for white people," they won't care much whether he delivers on specific promises. They feel represented and understood at last. *He is our voice.*

And he is their voice especially when it comes to religious identity. If anyone should be feeling dissonance about their support of Donald Trump, it is the voters for whom religious convictions are central to their self-concepts. The greater the dissonance caused by the gap between such a basic belief and support for a politician who violates virtually every ethical and moral element of that belief, the greater the need to either disavow the politician or justify his behavior. The choice has been clearest for white evangelical Christians, who made up 26 percent of voters in 2016: Fully 81 percent of them voted for Trump, and they remained steadfast in their support even after his fateful quid pro quo phone call with the Ukrainian president that launched the impeachment inquiry. The majority of evangelicals and Republicans who watch Fox News said in a survey there was nothing Trump could do to lose their approval and that nothing he has done has "hurt the dignity of the presidency."[35] How do they maintain this belief?

Before Donald Trump became the Republican nominee for president, Wanda Alger, who describes herself as a "prophetic minister," wrote a column for the conservative Christian publication *Charisma News* (tagline: "Where Faith and Politics Meet") titled "We Need the Fear of the Lord in Our Leaders."

"There is one prayer that this nation needs now more than ever before," she wrote. "It is the fear of the Lord that needs to grip the hearts and minds of those in public office as well as those who are voting." She listed sixteen qualities that should be apparent in all leaders "whose hearts are truly poised toward God," including these:

- They will be open to the counsel of others (Prov. 1:7)
- They will be teachable (Ps. 25:12)
- They will show no partiality and take no bribes (Deut. 10:17)
- They will not consider themselves better than anyone else (Deut. 17:19)
- Their mouths will be filled with good things (Prov. 16:9–13)
- They will hate all forms of evil (Prov. 8:13)
- They will operate in wisdom and humility (Prov. 15:33)
- Their house will be in order (Ps. 128:1–4)
- They will walk in obedience to God's commands (Ps. 86:11)
- They will let no sin rule over them (Ps. 119:133)
- They will have no fear of man, but walk in the fear of the Lord (Prov. 29:25)
- They will rule with justice, counsel, and might (Is. 11:2–4)

What is a good Christian to do with the dissonance generated by learning that Trump is in no way "teachable" or "open to counsel of others," that he takes bribes in the form of deals

362 MISTAKES WERE MADE (BUT NOT BY *ME*)

that benefit Trump holdings, that he rarely if ever operates in "wisdom and humility," rarely speaks "good things," barely keeps his house (the White one) in order, and, far from being humble and not considering himself "better than everyone else," freely repeats that he sees himself as better and smarter than everyone else? "No one respects women more than me," he said. "No one reads the Bible more than me." "There's nobody that's done so much for equality as I have." "Nobody knows more about taxes than me, maybe in the history of the world." "Nobody's ever been more successful than me."[36] He proclaimed himself "the Chosen One," a man of "great and unmatched wisdom."

There you are, a devout Christian, at the peak of the pyramid. This guy Trump is hardly a man of strong moral character. There is that pesky history of marital infidelity and vulgarity toward women, and after all, you were furious with Bill Clinton for far less. Trump has shown no history of commitment to, let alone belief in, Christian faith, despite all that Bible-reading he claims to have done. Do you say, "Trump sure doesn't live up to our requirements for a godly leader, so we better look elsewhere"? More likely you will take the route of dissonance reduction that Wanda Alger did. After watching Trump in action for two years, Alger miraculously realized that a good leader does not have to be godly after all. It would be a nice bonus if he is, she decided, but it doesn't matter, and moreover, it *never* mattered (apparently, she forgot that before the election, it mattered enormously to her). The only question people should be asking, she wrote in 2019, was whether Trump is a good *leader:* "For those leaders, including the president, whose demeanor or style may not fit our desired approach as Christians, we can look to the primary qualifications in Scripture to determine their ability to govern well. Unlike church leaders who must model Christ to the flock, civil government leaders are called to rule with a strong hand to ensure safety, protection and freedom for all."

And what are the "primary qualifications in Scripture" that make a good leader? Romans 13:1–6 and 1 Peter 2:13–14, she explained, describe the kind of leader we need:

- God's servant for the good of the people.
- A terror to bad conduct, bearing the sword.
- Avengers who carry out God's wrath on the wrongdoer.
- Servants of God, attending to taxation of the people.
- Sent by God to punish those who do evil and praise those who do good.[37]

This list is more Trumpian for sure, especially that bearing-the-sword part and the avenging part and the taxation part, though let's ignore the verses showing that Jesus cared about taxing the rich for the benefit of the poor, not the other way around (Mark 10:21, 25; Proverbs 19:17). "These descriptions of how God will use civil leaders include no indications of personal morality or godliness being necessary," Alger wrote. "Though it may be desired, it is not required. We must continue to pray that all our elected officials would have an authentic encounter and personal relationship with Jesus Christ. But let's not disqualify those who have not yet heard, or are still on the journey. If they are truly fulfilling God's purposes as civil leaders, their actions will speak louder than their words, and their accomplishments outweigh their personal weaknesses."

Translation: "We like his policies and his attitudes so much that we will abandon our moral values to support him and overlook his many sins while retaining our belief that we are good, kind, compassionate, devout Christians." Achieving that balance requires some fancy mental footwork. So great is the dissonance between Christian values and Trump's behavior that his evangelical supporters must work hard to minimize the criticism against him, especially, said Alger, from all those "leftist liberals who oppose religious freedom and conservative values [and who] are incessant in finding a means to disqualify our current president. Whether it's his negative comments on Twitter or his unorthodox methods in governing, his opponents are looking for anything to suggest he is not fit to govern the nation."

None of that criticism matters anyway, these evangelicals have concluded, because the ends justify the means, and God has sent Trump to give us the ends we want, not only by putting conservative judges in the courts but also by supporting Christian universities and organizations that object to same-sex marriage or contraception, permitting religious groups to be freed from anti-discrimination laws, and moving the U.S. embassy in Israel to Jerusalem. The ultimate end is that Trump will restore the United States to what they believe it was and should be — a white Christian nation. And to do that, he will block the onslaught of those "people of color" and non-Christian people and nonheterosexual people and foreigners who are invading our country, even if those foreigners were born in New York or Cincinnati. As John Fea, the author of *Believe Me: The Evangelical Road to Donald Trump,* put it: Evangelicals will "look away from the moral indiscretion in order to get their political agenda in place."

Where in history have we seen this tradeoff before? Ask Pius XI.

One who has certainly looked away is Mike Pompeo, an evangelical Christian who became secretary of state after Trump fired Rex Tillerson. Before the election, Pompeo, a conservative internationalist who regarded American power as crucial to global stability, was a vocal opponent of Trump's America First policies. Yet when offered a job that required him to sacrifice his political views and, presumably, his religious ones, he did so in a heartbeat, explaining that a secretary of state's job was to serve the president and do whatever the president asks of him. A former White House official said that Pompeo was "among the most sycophantic and obsequious people around Trump." Even more bluntly, a former American ambassador said, "He's like a heat-seeking missile for Trump's ass." And how does Pompeo justify this behavior? God raised up Trump — as God did with Queen Esther, who saved her people — to this exalted position. Therefore God, and Trump, must be obeyed.[38] Even if that means violating your oath of office and stonewalling Congress's requests for information pertaining to its impeachment inquiry.

Nonetheless, there is one of the Ten Commandments whose violation even evangelicals cannot abide in Trump. It's not adultery, of course, and it's not the prohibition against giving false witness, though the man made more than fifteen thousand false or misleading statements in the first three years of his presidency. It's not coveting thy neighbor's property either; that's just good business. The bridge too far was this: "Thou shalt not take the Lord's name in vain." Thus, when Trump joked about "goddamn windmills" while talking with House Republicans about energy policy, he infuriated many of his evangelical supporters. "I certainly do not condone taking the Lord's name in vain. There is a whole commandment dedicated to prohibiting that," said the Reverend Robert Jeffress, an evangelical who advises Trump and is one of his staunchest supporters. "I think it's very offensive to use the Lord's name in vain. I can take just about everything else, except that."[39] Never mind the other "whole commandments" Trump flouts daily.

In one of the most extreme forms of dissonance reduction, many of Trump's evangelical supporters do not merely excuse or minimize Trump's adulteries and deceits; they take them as evidence that they were right all along to support him. In fact, the more vulgar he is, the more he fulfills his supposedly divine mission. Hence one bumper sticker we saw: DONALD IS MINE/CHOSEN DIVINE. He literally *is* God or at least was sent by God to save us. Evangelicals don't see their endorsement of Trump as hypocrisy, much less heresy. "They believe that Trump is appointed by God for a moment such as this," says Fea. "They believe that God uses corrupt people — there are examples in the Bible of this, so they'll call upon these verses." After all, God works in mysterious ways, and no one's ways are more mysterious than Donald Trump's. Does he have a long litany of sins, sexual and financial? Well, God loves a sinner. Trump is on a "journey" to Christ, they say. Might take him a while to get there, but we're Christians; we're prepared to tolerate a few sinners — as long as they are our sinners. Some evangelicals, Fea says, go further, claiming that Trump has *already had* a spiritual awakening and that his days (and nights) of corruption are far in the past. "Donald Trump has changed," said retiree Nancy Allen, a Baptist from North Carolina who wrote *Electing the People's President, Donald Trump.* "I believe that with all my heart. He has changed. He hasn't had any more affairs. Now he's not perfect, but there's no perfect person. We know that there has been a change in his heart, and he respects our beliefs and values. And I believe he has some of the same beliefs and values."

At a 2019 conference of the Faith and Freedom Coalition, the group's chairman, Ralph Reed, told the cheering crowd: "We have had some great leaders. There has never been anyone who has defended us and fought for us, who we have loved more than Donald J. Trump. We have seen his heart and he is everything he promised he would be, and more."[40]

In an echo of Mussolini's boast to the German foreign minister about how easy it had been to win the Vatican's support, Trump, talking to GOP lawmakers, referred to "those fucking evangelicals" while smiling and shaking his head. In Trump's mind, writes Tim Alberta in *American Carnage,* he would "give them the policies and the access to authority that they longed for. In return they would stand behind him unwaveringly."

Stopping the Slide: "Look, This Is Bigger Than the Politics of the Day"

My loyalty to Mr. Trump has cost me everything — my family's happiness, friendships, my law license, my company, my livelihood, my honor, my reputation, and, soon, my freedom. I pray the country doesn't make the same mistakes I have made.

— *Michael Cohen, Donald Trump's former personal attorney and fixer, on being sentenced to three years in prison*

Michael Cohen was no choirboy. He didn't have much in common with Jeb Stuart Magruder, whose "ethical compass" would almost certainly have been shattered by many of Cohen's lifelong, sleazy professional dealings both before and after his association with Donald Trump. But when Cohen was caught, he saw what his blind loyalty had cost him; he decided to cop a plea and take the rap. He had evaded taxes, but he could not evade the consequences of unquestioned loyalty to a con man.

For most of Cohen's life, the ends justified the means — until they couldn't. It's a choice many of us have to make in our lives, whether the goals we seek are small or large, personal or political. We often have to determine whether a specific goal we care about — especially a righteous one, such as justice for abuse victims, ending sexual harassment in the workplace, or achieving a particular political reform — is more important than what we do to reach it. So what, we say, if we have to make a few unsavory alliances to get there? So what if a few innocent people are thrown under the bus? Certainly Pope Pius XI learned the answer: Sooner or later, we are likely to be thrown under that bus ourselves... or the bus will careen off the road with everyone in it.

Throughout this book, we have seen why most people, when faced with dissonance between their ambitions and their ethics, step off the pyramid of choice in the direction of ambition — nudged over by convenience, peer support, job security, and other rewards — swallowing their doubts and letting self-justification ease their consciences. We have ended most chapters with a story of someone who took the harder route, and for us, those stories reveal not only individual courage but also the powerful web that dissonance reduction weaves to keep us in line.

Consider again the trajectory down the pyramid that began with the Never Trump Republicans, a group that coalesced before the 2016 election. Some, such as George Will, remained adamant; for him, *never* meant *never*. Most, as we saw, eventually became supporters and slid to the bottom with full acquiescence to the administration. Others dug in their heels and stopped their slide down when they finally reached a breaking point, the point at which they could no longer justify their continued loyalty.

For Max Boot, disillusionment with his party was "painful and prolonged; in fact, existential." In his 2018 book *The Corrosion of Conservatism,* he argued that it was time for the Republican Party to pay "for its embrace of white nationalism and know-nothingism." For this to happen, he wrote, "the G.O.P. as currently constituted [must be] burned to the ground."[41]

For Jim Mattis, the former secretary of defense, the last straw was Trump's abrupt announcement at the end of 2018 that he was withdrawing American forces in Syria, where they were fighting the Islamic State. This move would mean abandoning America's Kurdish allies and giving Turkey and Russia the political plum that they wanted. Mattis, a strong believer in alliances, knew that a retreat from Syria would threaten the security of American troops elsewhere in the region as well as infuriate the Kurds and other allies in the anti-ISIS coalition, who would justifiably feel betrayed. Mattis urged Trump to reconsider his decision, but Trump remained obstinate. Mattis was willing to stay on another two months to minimize disruption in the department. Trump, who cannot abide dissent for even a week, fired him a few days later.*

For the influential neoconservative William Kristol, it was the unwillingness of the Republican Party to hold Trump accountable for the findings in Mueller's 2019 report to Congress. He and other like-minded organizers formed Republicans for the Rule of Law. A spokesman for the group told *Newsweek*, "Everybody — Republicans and Democrats but especially Republicans — needs to step up and say, 'Look, this is bigger than the politics of the day, this is about our democratic institutions.' If we don't defend them, that will have an impact on our country for decades to come. President Trump still does not want to

---

* Trump modified that position almost as soon as Mattis departed; he left a few hundred troops in place, but in October of 2019, without consulting his military experts at the Pentagon, he impulsively withdrew all American troops — and Mattis's prediction of disaster immediately came to pass. Abandoning the Kurds was not only an immoral betrayal of a key ally in the fight against ISIS but an act that had devastating military and political ramifications. Even Trump's stalwart Republican supporters were furious; two-thirds of House Republicans joined Democrats in approving a resolution that opposed Trump's decision.

admit that this happened and that's wrong, absurd and danger-ous. Republicans need to stop enabling this behavior."

For Justin Amash, the first Republican member of Congress to call for Trump's impeachment, it was reading the Mueller Report. Amash announced on July 4, 2019, that he was declaring his own independence and leaving the Republican Party.

For the conservative columnist Peter Wehner, who served under three Republican presidents, the struggle to "balance the scales of [Trump's] conservative achievements (like with the courts) against the harm he's caused and the ways he's changed the Republican Party and the country"[42] finally resolved itself: harm far outweighed achievements. "Trump has shown himself to be a pathological liar engaged in an all-out assault on objective facts — on reality and truth — concepts on which self-govern-ment depends," he wrote. "The president is also cruel, and dehu-manizes his opponents. He's volatile and emotionally unstable. He relishes dividing Americans along racial and ethnic lines. He crashes through norms like a drunk driver crashes through guardrails. And he's corrupt from stem to stern."

We do not underestimate how difficult it is for anyone to shed a political affiliation that, as Max Boot put it, is "existential," defining one's values, self-concept, and worldview. Nevertheless, as ethical as these Republicans have been, their repudiation of Trump did not threaten their livelihoods. For others, peeling away from the pack was more difficult and the price higher.

Shane Claiborne is an evangelical who preaches the gospel, befriends prisoners on death row, makes his own clothes, and lives among the poor. But when he came to Lynchburg, Virginia, in 2018 to preach at a Christian revival at Jerry Falwell's Liberty

University, the chief of police sent him a warning: If he set foot on the property, he would be arrested for trespassing and face up to twelve months in jail and a twenty-five-hundred-dollar fine. What was Shane Claiborne planning to do that evoked such an un-Christian response? Claiborne and a small but vocal group of liberal evangelicals wanted to go "where toxic Christianity lives," preaching their moral and theological objections to the evangelical majority's support of Donald Trump. They are angry that their church has been endorsing Trump's program of deporting immigrants, fanning racial tension, reducing help for the poor, and passing a tax cut for the rich. "There is another Gospel in our country right now," Claiborne exhorted the audiences attending his sermons, "and it is the Gospel of Trump. It doesn't look much like the Gospel of Jesus." This sentiment was echoed by Ben Howe, who was raised in a strict evangelical family, in his book *The Immoral Majority: Why Evangelicals Chose Political Power over Christian Values.* "We need to own our own mistakes and change our own Christian subculture," he wrote, "if we want those who currently see evangelicals as unpersuasive hypocrites ever to be open to listening again. But so far, Christians have not shown an eagerness to rise to that challenge; they seem much more prepared to shed their principles for Donald Trump than they are willing to work to earn the trust of a disenchanted public."[43]

Establishment evangelicals have not welcomed these dissenters. Lynchburg is a company town in which Liberty University is the biggest employer, and no one wants to cross Jerry Falwell Jr. When reporter Laurie Goodstein went to Lynchburg to interview local pastors, most of those who supported the dissenters said they would nonetheless remain silent for fear of displeasing their congregations and risking their jobs. "Everyone's afraid," one minister told her. "That's strong language. Everyone's very mindful of how they speak and how they deliver the truth. It's hard to tell the truth in a context like Lynchburg."[44]

Maria Caffrey, a climate scientist who worked at the National Park Service Natural Resource Stewardship and Science Directorate, eventually lost her job for refusing to hide the facts — the factual facts, not the alternative ones — that explained the crisis of climate change. "Senior NPS [National Park Service] officials tried repeatedly, often aggressively, to coerce me into deleting references to the human causes of the climate crisis," she wrote. "This was not normal editorial adjustment. This was climate science denial. . . . They threatened to make the deletions without my approval if I would not agree, to release the report without naming me as the primary author, or not release it all. Each option would have been devastating to my career and for scientific integrity. I stood firm."[45] At first, Caffrey prevailed by getting the word to the media and members of Congress, and her report was published as she had written it. But NPS higher-ups continued to retaliate against her with pay cuts, demotions, and, ultimately, refusing to renew funding for the projects she had created and was working on. Her colleagues advised her to quit and become a volunteer. She did. Her application to become a volunteer was denied.

Caffrey did her best, and she was fired. Most bite their tongues, hunker down, and try to do their jobs as best they can until the time comes when they can't live with themselves any longer. Chuck Park, a U.S. Foreign Service officer, son of South Korean immigrants, found himself in a constant state of dissonance, "struggling to explain to foreign peoples the blatant contradictions at home." Every day, he wrote in an op-ed, he found it harder to refuse visas based on administration priorities, recite administration talking points on border security, and support Trump appointees who pushed his "toxic agenda" around the world. Facing his dissonance, he articulated its conflicting elements: As a Foreign Service officer, he was obligated to serve "during the pleasure of the President of the United States" and follow the administration's "pleasure" — or else quit. "I let career perks silence my conscience," he said. "I let free housing, the countdown to a pension and the prestige of representing a powerful nation overseas distract me from ideals that once seemed so clear to me. I can't do that anymore. My son, born in El Paso . . . the same city where 22 people were just killed by a gunman whose purported 'manifesto' echoed the inflammatory language of our president, turned 7 this month. I can no longer justify to him, or to myself, my complicity in the actions of this administration. That's why I choose to resign."[46]

And for some, such as that first whistleblower in the CIA who formally accused Donald Trump of wrongdoing, the stakes of resolving dissonance on the side of principles and patriotism are even higher. Social scientists who have studied the psychology of whistleblowing know how perilous this action can be. Americans *say* that they value the courage of those rare employees who alert the public to safety violations, crimes, and unethical behavior on the part of their employers. But most whistleblowers end up paying a big price; they often lose their jobs, families, friends, and security. Knowing this, in addition to knowing Trump's fear-generating claims that whistleblowers are committing "treason," the unnamed (as of this writing) intelligence officer who chose to file an official complaint, even though it was done within the prescribed procedures, acted with immense courage.

So, most assuredly, did the ousted Ukraine ambassador Marie L. Yovanovitch, a thirty-three-year veteran of the State Department who served under six presidents, both Republican and Democratic, when she defied the administration's order not to appear for the House's impeachment proceedings. Although her superiors had told her she had "done nothing wrong" — that, on the contrary, her knowledge and experience in Ukraine had proved invaluable — Donald Trump wanted her removed because her anti-corruption policy was blocking the efforts of the president's personal lawyer Rudolph W. Giuliani and two of his associates to try to find damaging information about Biden. (The two associates were subsequently indicted for campaign-finance violations, among other charges; they were arrested as they boarded an international flight holding one-way tickets.) These men, she said, passed along what she called "scurrilous lies" about her. "Although I understand that I served at the pleasure of the president, I was nevertheless incredulous that the U.S. government chose to remove an ambassador based, as best as I can tell, on unfounded and false claims by people with clearly questionable motives," she said in her statement to Congress. The State Department is "becoming hollowed out from within," she warned.

> The harm will come not just through the inevitable and continuing resignation and loss of many of this nation's most loyal and talented public servants. It also will come

when those diplomats who soldier on and do their best to represent our nation face partners abroad who question whether the ambassador truly speaks for the president and can be counted upon as a reliable partner. The harm will come when private interests circumvent professional diplomats for their own gain, not the public good.[47]

The floodgates were opened. Before long, other distinguished and experienced public servants testified before the House Intelligence Committee and confirmed her account, one after another doing so, they said, out of a sense of duty.

• • •

This has been a book about how difficult it is to own our mistakes and the crucial importance of doing so if we ever hope to learn and improve. Millions of citizens made a monumental mistake in electing, and then supporting, Donald Trump. Once he is gone from the scene, the self-forgiving distortions of memory will lead many of his former supporters to say, "I never voted for him anyway" or "I always had misgivings about him." Many of his former opponents will breathe a sigh of relief and say, "Thank God that's over." But we can never be complacent again. We all need to stand back and ask: What have we learned?

We have learned how precarious democracy is, how easily fear and anger can be invoked to manipulate a population. We have learned about the importance of voting, even if it means choosing a candidate we regard as the lesser of two evils rather than our number-one purely pristine perfect preference. We have learned that a democracy rests not only on its laws and institutions but also on its norms and values — and on the consensus of its citizenry that those norms and values are worth upholding. We have learned that heeding the rules of civility, decency, and diplomacy is a sign not of a nation's weakness but of its strength.

Donald Trump has broken the rules and norms of democracy with flagrant arrogance, but liberals and conservatives alike have observed that in so doing, he has forced us to pay heed to our nation's vulnerability and determine what kind of a country we want to be. From the left, David Remnick, editor of the *New Yorker,* wrote, "It's entirely possible that Donald Trump, who has been such a ruinous figure on the public scene, has at least done the country an unintended service by clarifying some of our deepest flaws and looming dangers in his uniquely lurid light."[48] And from the right, General Jim Mattis, Trump's former secretary of defense, wrote that "all Americans need to recognize that our democracy is an experiment — and one that can be reversed. We all know that we're better than our current politics. Tribalism must not be allowed to destroy our experiment."[49]

In the final analysis, Republicans, Democrats, and independents who are fearful of what Trump has done to the moral fabric of our country will not find the way forward clear or easy. People are tired of arguing with their brothers-in-law, and many are tired, period. But there is too much at stake to turn away. By understanding the mechanisms that keep people wedded to their initial justifications of a decision, citizens can — with insight and a willingness to admit error — get our country back on course. Donald Trump does not learn from his mistakes, but we remain hopeful that our nation can.

ACKNOWLEDGMENTS

We decided the order of authorship of this book by flipping a coin; it's that balanced a collaboration. However, from start to finish, each of us has firmly believed that he or she was working with the more talented coauthor. So, to begin with, we want to thank each other for making this project one of mutual encouragement and learning—and fun.

Our book has benefited from careful, critical readings by colleagues who are specialists in the areas of memory, law, couples therapy, business, and clinical research and practice. We would especially like to thank the following colleagues for their close evaluation of chapters in their fields of expertise, and for the many excellent suggestions they gave us: Andrew Christensen, Deborah Davis, Gerald Davison, Maryanne Garry, Samuel Gross, Bruce Hay, Brad Heil, Richard Leo, Scott Lilienfeld, Elizabeth Loftus, Andrew McClurg, Devon Polachek, Donald Saposnek, and Leonore Tiefer. In addition, we appreciate the comments, ideas, stories, research, and other information offered by J. J. Cohn, Joseph de Rivera, Ralph Haber, Robert Kardon, Saul Kassin, Burt Nanus, Debra Poole, Anthony Pratkanis, Holly Stocking, and Michael Zagor. Our thanks also to Deborah Cady and Caryl McColly for their editorial help.

Our editors and production team have consistently been superb. In the original edition, we thanked our commissioning editor, Jane Isay, whose stories and ideas infuse this book and who has remained a staunch supporter and adviser over subsequent revisions; supervising editor Jenna Johnson; managing editor David Hough; and Margaret Jones, for exceptional copyediting and fact checking. For the updated edition, we added our thanks and appreciation to editorial director Ken Carpenter; book designers Christopher Moisan, Greta Sibley, and Chrissy Kurpeski; Tim Mudie, our hands-on editor; and our thorough, witty copyeditor Tracy Roe, who created a certain amount of dissonance in us by catching more mistakes than we should have made. Thank goodness she wasn't bored with us and was willing to take us on for this latest edition with her usual eagle eye, wry notes, and helpful suggestions.

For this edition, we also wish to express our gratitude to editorial manager Nicole Angeloro, for enthusiastically giving us the opportunity to update this book to deal with the important issues of our times; senior production editor Lisa Glover, for her patient and meticulous shepherding of the book through a tight production schedule; Emily Snyder, for so creatively matching yet updating the book's design; Michael Dudding, for expertly handling the marketing of our book to diverse audiences; and the rest of the excellent production team at Mariner Books.

Carol wishes to honor the memory of Ronan O'Casey for his love and support in their many years together; Elliot, in his signature phrase, gives, "of course," his love and thanks to Vera Aronson. Mistakes were made by us in our lives, but not in the choice of a life partner.

— *Carol Tavris and Elliot Aronson*

NOTES

Long before we became writers, we were readers. As readers, we often found notes an unwelcome intrusion in the flow of the story. It was usually a pain in the neck to be forever turning to the back of the book to learn what the author's source was for some persuasive (or preposterous) idea or research finding, but every so often there was candy — a personal comment, an interesting digression, a good story. We enjoyed assembling these notes, using the opportunity to reference and sometimes expand the points we make in the chapters. And there's some candy in here too.

INTRODUCTION

1. "Spy Agencies Say Iraq War Worsens Terrorism Threat," *New York Times,* September 24, 2006; the comment to conservative columnists was reported by one of them, Max Boot, in "No Room for Doubt in the Oval Office," the *Los Angeles Times,* September 20, 2006. For a detailed accounting of George Bush's claims to the public regarding the war in Iraq, see Frank Rich, *The Greatest Story Ever Sold: The Decline and Fall of Truth from 9/11 to Katrina* (New York: Penguin, 2006). In his State of the

Union address in January 2007, Bush acknowledged that "where mistakes were made" in a few tactics used in conducting the war, he was responsible for them. But he held firm that there would be no major changes in strategy; on the contrary, he would increase the number of troops and invest even more money in the war. In their memoirs, high-ranking members of Bush's administration have painted a portrait of Bush and his inner circle as people driven by certainty and "groupthink"; anyone who offered unwelcome facts was ignored, demoted, or fired. See, for example, Robert Draper, *Dead Certain: The Presidency of George W. Bush* (New York: Free Press, 2007); Jack Goldsmith, *The Terror Presidency: Law and Judgment Inside the Bush Administration* (New York: W. W. Norton, 2007); and Michael J. Mazarr, *Leap of Faith: Hubris, Negligence, and American's Greatest Foreign Policy Tragedy* (New York: PublicAffairs, 2019). Within one day of 9/11, Mazarr writes, the decision to overthrow Saddam Hussein "had been essentially sealed in cognitive amber."

2. The American Presidency Project (online), www.presidency.ucsb.edu/ws/index.php, provides documented examples of every instance of "mistakes were made" said by American presidents. It's a long list. Bill Clinton said that "mistakes were made" in the pursuit of Democratic campaign contributions and later joked about the popularity of this phrase and its passive voice at a White House Press Correspondents dinner. Of all the presidents, Richard Nixon and Ronald Reagan used the phrase most, the former to minimize the illegal actions of the Watergate scandal, the latter to minimize the illegal actions of the Iran-Contra scandal. See also Charles Baxter's eloquent essay "Dysfunctional Narratives, or: 'Mistakes Were Made,'" in Charles Baxter, *Burning Down the House: Essays on Fiction* (Saint Paul, MN: Graywolf Press, 1997).

3. Gordon Marino, "Before Teaching Ethics, Stop Kidding Yourself," *Chronicle of Higher Education* (February 20, 2004): B5.

4. On the self-serving bias in memory (and the housework study in particular), see Michael Ross and Fiore Sicoly, "Egocentric Biases in Availability and Attribution," *Journal of Personality and Social Psychology* 37 (1979): 322–36. See also Suzanne C. Thompson and

Harold H. Kelley, "Judgments of Responsibility for Activities in Close Relationships," *Journal of Personality and Social Psychology* 41 (1981): 469–77.

5. John Dean, interview by Barbara Cady, *Playboy,* January 1975, 78.

6. Robert A. Caro, *Master of the Senate: The Years of Lyndon Johnson* (New York: Knopf, 2002), 886.

7. Katherine S. Mangan, "A Brush with a New Life," *Chronicle of Higher Education* (April 2005): A28–A30.

8. Sherwin Nuland, *The Doctors' Plague: Germs, Childbed Fever, and the Strange Story of Ignác Semmelweis* (New York: Norton, 2003).

9. Ferdinand Lundberg and Marynia F. Farnham, *Modern Woman: The Lost Sex* (New York: Harper and Brothers, 1947), 11, 120.

10. Edward Humes, *Mean Justice* (New York: Pocket Books, 1999).

1. COGNITIVE DISSONANCE: THE ENGINE OF SELF-JUSTIFICATION

1. Press releases from Neal Chase, representing the religious group Baha'is Under the Provisions of the Covenant, in "The End Is Nearish," *Harper's,* February 1995, 22, 24.

2. Leon Festinger, Henry W. Riecken, and Stanley Schachter, *When Prophecy Fails* (Minneapolis: University of Minnesota Press, 1956).

3. O. Fotuhi et al., "Patterns of Cognitive Dissonance-Reducing Beliefs Among Smokers: A Longitudinal Analysis from the International Tobacco Control (ITC) Four Country Survey," *Tobacco Control: An International Journal* 22 (2013): 52–58; and F. Naughton, H. Eborall, and S. Sutton, "Dissonance and Disengagement in Pregnant Smokers," *Journal of Smoking Cessation* 8 (2012): 24–32.

4. Leon Festinger, *A Theory of Cognitive Dissonance* (Stanford, CA: Stanford University Press, 1957). See also Leon Festinger and Elliot Aronson, "Arousal and Reduction of Dissonance in Social Contexts," in *Group Dynamics,* eds. D. Cartwright and Z. Zander (New York: Harper and Row, 1960–61); and Eddie Harmon-Jones and Judson Mills, eds., *Cognitive Dissonance: Progress on a Pivotal Theory in Social Psychology* (Washington, DC: American Psychological Association, 1999).

5. Elliot Aronson and Judson Mills, "The Effect of Severity of Initiation on Liking for a Group," *Journal of Abnormal and Social Psychology* 59 (1959): 177–81.

6. Harold Gerard and Grover Mathewson, "The Effects of Severity of Initiation on Liking for a Group: A Replication," *Journal of Experimental Social Psychology* 2 (1966): 278–87.

7. Dimitris Xygalatas et al., "Extreme Rituals Promote Prosociality," *Psychological Science* 24 (2013): 1602–5.

8. Many cognitive psychologists and other scientists have written about the confirmation bias. See Thomas Kida, *Don't Believe Everything You Think* (Amherst, NY: Prometheus Press, 2006), and Raymond S. Nickerson, "Confirmation Bias: A Ubiquitous Phenomenon in Many Guises," *Review of General Psychology* 2 (1998): 175–220.

9. Claudia Fritz et al., "Soloist Evaluations of Six Old Italian and Six New Violins," *Proceedings of the National Academy of Sciences* 111 (2014): 7224–29, doi: 10.1073/pnas.1323367111.

10. Adrian Cho, "Million-Dollar Strads Fall to Modern Violins in Blind 'Sound Check,'" ScienceMag.org, May 9, 2017.

11. Lenny Bruce, *How to Talk Dirty and Influence People* (Chicago: Playboy Press, 1966), 232–33.

12. Steven Kull, director of the Program on International Policy Attitudes (PIPA) at the University of Maryland, commenting on the results of the PIPA/Knowledge Networks poll "Many Americans Unaware WMD Have Not Been Found," June 14, 2003.

13. Gary C. Jacobson, "Perception, Memory, and Partisan Polarization on the Iraq War," *Political Science Quarterly* 125 (Spring 2010): 1–26. See also his paper "Referendum: The 2006 Midterm Congressional Elections," *Political Science Quarterly* 122 (Spring 2007): 1–24.

14. Drew Westen et al., "The Neural Basis of Motivated Reasoning: An fMRI Study of Emotional Constraints on Political Judgment During the U.S. Presidential Election of 2004," *Journal of Cognitive Neuroscience* 18 (2006): 1947–58. For readers interested in the neuroscience of dissonance, see also Eddie Harmon-Jones, Cindy Harmon-Jones, and

David M. Amodio, "A Neuroscientific Perspective on Dissonance, Guided by the Action-Based Model," in *Cognitive Consistency: A Fundamental Principle in Social Cognition,* eds. B. Gawronski and F. Strack (New York: Guilford, 2012), 47–65; and S. Kitayama et al., "Neural Mechanisms of Dissonance: An fMRI Investigation of Choice Justification," *NeuroImage* 69 (2013): 206–12.

15. Charles Lord, Lee Ross, and Mark Lepper, "Biased Assimilation and Attitude Polarization: The Effects of Prior Theories on Subsequently Considered Evidence," *Journal of Personality and Social Psychology* 37 (1979): 2098–2109.

16. Brendan Nyhan and Jason Reifler, "When Corrections Fail: The Persistence of Political Misperceptions," *Political Behavior* 32 (2010): 303–30; Stephan Lewandowsky et al., "Misinformation and Its Correction: Continued Influence and Successful Debiasing," *Psychological Science in the Public Interest* 13 (2012): 106–31.

17. Doris Kearns Goodwin, *No Ordinary Time* (New York: Simon and Schuster, 1994), 321. (Emphasis in original.)

18. In one of the earliest demonstrations of postdecision dissonance reduction, Jack Brehm, posing as a marketing researcher, showed a group of women eight different appliances (a toaster, a coffeemaker, a sandwich grill, and the like) and asked them to rate each item for its desirability. Brehm then told each woman that she could have one of the appliances as a gift, and he gave her a choice between two of the products she had rated as being equally appealing. After she chose one, he wrapped it up and gave it to her. Later, the women rated the appliances again. This time, they increased their rating of the appliance they had chosen and decreased their rating of the appliance they had rejected. See Jack Brehm, "Postdecision Changes in the Desirability of Alternatives," *Journal of Abnormal and Social Psychology* 52 (1956): 384–89.

19. Robert E. Knox and James A. Inkster, "Postdecision Dissonance at Post Time," *Journal of Personality and Social Psychology* 8 (1968): 319–23. There have been many replications of the finding that the more permanent and less revocable the decision, the stronger the need to

reduce dissonance. See Lottie Bullens et al., "Reversible Decisions: The Grass Isn't Merely Greener on the Other Side; It's Also Very Brown Over Here," *Journal of Experimental Social Psychology* 49 (2013): 1093–99.

20. Katherine S. Mangan, "A Brush with a New Life," *Chronicle of Higher Education* (April 2005): A28–A30.

21. Brad J. Bushman, "Does Venting Anger Feed or Extinguish the Flame? Catharsis, Rumination, Distraction, Anger, and Aggressive Responding," *Personality and Social Psychology Bulletin* 28 (2002): 724–31; Brad J. Bushman et al., "Chewing on It Can Chew You Up: Effects of Rumination on Triggered Displaced Aggression," *Journal of Personality and Social Psychology* 88 (2002): 969–83. The history of research disputing the assumption of catharsis is summarized in Carol Tavris, *Anger: The Misunderstood Emotion* (New York: Simon and Schuster, 1989).

22. Michael Kahn's original study was "The Physiology of Catharsis," *Journal of Personality and Social Psychology* 3 (1966): 78–98. For another early classic, see Leonard Berkowitz, James A. Green, and Jacqueline R. Macaulay, "Hostility Catharsis as the Reduction of Emotional Tension," *Psychiatry* 25 (1962): 23–31.

23. Jon Jecker and David Landy, "Liking a Person as a Function of Doing Him a Favor," *Human Relations* 22 (1969): 371–78.

24. Nadia Chernyak and Tamar Kushnir, "Giving Preschoolers Choice Increases Sharing Behavior," *Psychological Science* 24 (2013): 1971–79.

25. Benjamin Franklin, *The Autobiography of Benjamin Franklin* (New York: Touchstone, 2004), 83–84.

26. Ruth Thibodeau and Elliot Aronson, "Taking a Closer Look: Reasserting the Role of the Self-Concept in Dissonance Theory," *Personality and Social Psychology Bulletin* 18 (1992): 591–602.

27. Jonathon D. Brown, "Understanding the Better than Average Effect: Motives (Still) Matter," *Personality and Social Psychology Bulletin* 38 (2012): 209–19. Brown showed in a series of experiments that the "I'm better than average" effect increases after a person experiences a threat to his or her self-worth.

28. There is a large and lively research literature on the self-serving bias, the tendency to believe the best of ourselves and explain away the worst. It is a remarkably consistent bias in human cognition, though there are interesting variations across cultures, ages, and genders. See Amy Mezulis et al., "Is There a Universal Positivity Bias in Attributions? A Meta-Analytic Review of Individual, Developmental, and Cultural Differences in the Self-Serving Attributional Bias," *Psychological Bulletin* 130 (2004): 711–47; and Keith E. Stanovich et al., "Myside Bias, Rational Thinking, and Intelligence," *Psychological Science* 22 (2013): 259–64.

29. Philip E. Tetlock, *Expert Political Judgment: How Good Is It? How Can We Know?* (Princeton, NJ: Princeton University Press, 2005). In clinical psychology, the picture is the same; there is an extensive scientific literature showing that behavioral, statistical, and other objective measures of behavior are consistently superior to the insight of experts and their clinical predictions and diagnoses. See Robin Dawes, David Faust, and Paul E. Meehl, "Clinical Versus Actuarial Judgment," *Science* 243 (1989): 1668–74; W. M. Grove and Paul E. Meehl, "Comparative Efficiency of Formal (Mechanical, Algorithmic) and Informal (Subjective, Impressionistic) Prediction Procedures: The Clinical/Statistical Controversy," *Psychology, Public Policy, and Law* 2 (1996): 293–323; and Daniel Kahneman, "The Surety of Fools," *New York Times Magazine,* October 23, 2011.

30. Josh Barro, "The Upshot: Sticking to Their Story: Inflation Hawks' Views Are Independent of Actual Monetary Outcomes," *New York Times,* October 2, 2014.

31. Elliot Aronson and J. Merrill Carlsmith, "Performance Expectancy as a Determinant of Actual Performance," *Journal of Abnormal and Social Psychology* 65 (1962): 178–82. See also William B. Swann Jr., "To Be Adored or to Be Known? The Interplay of Self-Enhancement and Self-Verification," in *Motivation and Cognition,* eds. R. M. Sorrentino and E. T. Higgins (New York: Guilford, 1990); and William B. Swann Jr., J. Gregory Hixon, and Chris de la Ronde, "Embracing the Bitter 'Truth':

Negative Self-Concepts and Marital Commitment," *Psychological Science* 3 (1992): 118–21.

32. We are not idly speculating here. In a classic experiment conducted half a century ago, social psychologist Judson Mills measured the attitudes of sixth-grade children toward cheating. He then had them participate in a competitive exam with prizes offered to the winners. He arranged the situation so that it was almost impossible for a child to win without cheating and also so the children thought that they could cheat without being detected. (He was secretly keeping an eye on them.) About half the kids cheated and half did not. The next day, Mills asked the children again how they felt about cheating and other misdemeanors. Those children who had cheated became more lenient toward cheating, and those who resisted the temptation adopted a harsher attitude. See Judson Mills, "Changes in Moral Attitudes Following Temptation," *Journal of Personality* 26 (1958): 517–31.

33. Vivian Yee, "Elite School Students Describe the How and Why of Cheating," *New York Times,* September 26, 2012; Jenna Wortham, "The Unrepentant Bootlegger," *New York Times,* September 27, 2014.

34. Jeb Stuart Magruder, *An American Life: One Man's Road to Watergate* (New York: Atheneum, 1974), 4, 7.

35. Ibid., 194–95, 214–15.

36. The number of total participants is an informed estimate from psychologist Thomas Blass, who has written extensively about the original Milgram experiment and its many successors. About eight hundred people participated in Milgram's own experiments; the rest were in replications or variations of the basic paradigm over a twenty-five-year span.

37. The original study is described in Stanley Milgram, "Behavioral Study of Obedience," *Journal of Abnormal and Social Psychology* 67 (1963): 371–78. Milgram reported his study in greater detail and with additional supporting research, including many replications, in his subsequent book, *Obedience to Authority: An Experimental View* (New York: Harper and Row, 1974).

38. William Safire, "Aesop's Fabled Fox," *New York Times,* December 29, 2003.

2. PRIDE AND PREJUDICE . . . AND OTHER BLIND SPOTS

1. James Bruggers, "Brain Damage Blamed on Solvent Use," *Louisville Courier-Journal,* May 13, 2001; James Bruggers, "Researchers' Ties to CSX Raise Concerns," *Courier-Journal,* October 20, 2001; Carol Tavris, "The High Cost of Skepticism," *Skeptical Inquirer* (July/August, 2002): 42–44; Stanley Berent, "Response to 'The High Cost of Skepticism,'" *Skeptical Inquirer* (November/December 2002), 61, 63, 64–65. On February 12, 2003, the Office for Human Research Protections wrote to the vice president for research at the University of Michigan noting that the university's institutional review board, of which Stanley Berent had been head, had "failed to document the specific criteria for waiver of informed consent" for Berent and Albers's research. The case of CSX, its arrangement with Stanley Berent and James Albers, and their conflict of interest is also described in depth in Sheldon Krimsky, *Science in the Private Interest* (Lanham, MD: Rowman and Littlefield, 2003), 152–53.

2. Joyce Ehrlinger, Thomas Gilovich, and Lee Ross, "Peering into the Bias Blind Spot: People's Assessments of Bias in Themselves and Others," *Personality and Social Psychology Bulletin* 31 (2005): 680–92; Emily Pronin, Daniel Y. Lin, and Lee Ross, "The Bias Blind Spot: Perceptions of Bias in Self versus Others," *Personality and Social Psychology Bulletin* 28 (2002): 369–81. Our blind spots also allow us to see ourselves as being smarter and more competent than most people, which is why all of us, apparently, feel we are above average. See David Dunning et al., "Why People Fail to Recognize Their Own Incompetence," *Current Directions in Psychological Science* 12 (2003): 83–87, and Joyce Ehrlinger et al., "Why the Unskilled Are Unaware: Further Explorations of (Absent) Self-Insight Among the Incompetent," *Organizational Behavior and Human Decision Processes* 105 (2008): 98–121.

3. Quoted in Eric Jaffe, "Peace in the Middle East May Be Impossible:

Lee D. Ross on Naive Realism and Conflict Resolution," *American Psychological Society Observer* 17 (2004): 9–11.

4. Geoffrey L. Cohen, "Party over Policy: The Dominating Impact of Group Influence on Political Beliefs," *Journal of Personality and Social Psychology* 85 (2003): 808–22. See also Donald Green, Bradley Palmquist, and Eric Schickler, *Partisan Hearts and Minds: Political Parties and the Social Identities of Voters* (New Haven, CT: Yale University Press, 2002). This book shows how once people form a political identity, usually in young adulthood, the identity does their thinking for them. That is, most people do not choose a party because it reflects their views; rather, once they choose a party, its policies become their views.

5. Lee Epstein, Christopher M. Parker, and Jeffrey A. Segal, "Do Justices Defend the Speech They Hate?," paper presented at the 2013 annual meeting of the American Political Science Association, http://ssrn.com/abstract=2300572.

6. Emily Pronin, Thomas Gilovich, and Lee Ross, "Objectivity in the Eye of the Beholder: Divergent Perceptions of Bias in Self versus Others," *Psychological Review* 111 (2004): 781–99.

7. When privilege is a result of birth or another fluke of fortune rather than merit, many of its possessors will justify it as being earned. John Jost and his colleagues have been studying the processes of system justification, a psychological motive to defend and justify the status quo; see, for example, John Jost and Orsolya Hunyady, "Antecedents and Consequences of System-Justifying Ideologies," *Current Directions in Psychological Science* 14 (2005): 260–65. One such system-justifying ideology is that the poor may be poor but they are happier and more honest than the rich; see Aaron C. Kay and John T. Jost, "Complementary Justice: Effects of 'Poor but Happy' and 'Poor but Honest' Stereotype Exemplars on System Justification and Implicit Activation of the Justice Motive," *Journal of Personality and Social Psychology* 85 (2003): 823–37. See also Stephanie M. Wildman, ed., *Privilege Revealed: How Invisible Preference Undermines America* (New York: New York University Press, 1996).

8. D. Michael Risinger and Jeffrey L. Loop, "Three Card Monte, Monty Hall, Modus Operandi and 'Offender Profiling': Some Lessons of Modern Cognitive Science for the Law of Evidence," *Cardozo Law Review* 24 (November 2002): 193.

9. Dorothy Samuels, "Tripping Up on Trips: Judges Love Junkets as Much as Tom DeLay Does," *New York Times,* January 20, 2006.

10. Melody Petersen, "A Conversation with Sheldon Krimsky: Uncoupling Campus and Company," *New York Times,* September 23, 2003. Krimsky also recounted the Jonas Salk remark.

11. Krimsky, *Science in the Private Interest;* Sheila Slaughter and Larry L. Leslie, *Academic Capitalism* (Baltimore: Johns Hopkins University Press, 1997); Derek Bok, *Universities in the Marketplace: The Commercialization of Higher Education* (Princeton, NJ: Princeton University Press, 2003); Marcia Angell, *The Truth about the Drug Companies* (New York: Random House, 2004); and Jerome P. Kassirer, *On the Take: How Medicine's Complicity with Big Business Can Endanger Your Health* (New York: Oxford University Press, 2005).

12. National Institutes of Health Care Management Research and Educational Foundation, "Changing Patterns of Pharmaceutical Innovation," cited in Jason Dana and George Loewenstein, "A Social Science Perspective on Gifts to Physicians from Industry," *Journal of the American Medical Association* 290 (2003): 252–55.

13. Investigative journalist David Willman won a Pulitzer Prize for his series on conflicts of interest in bringing new drugs to market; two of them include "Scientists Who Judged Pill Safety Received Fees," *Los Angeles Times,* October 29, 1999; and "The New FDA: How a New Policy Led to Seven Deadly Drugs," *Los Angeles Times,* December 20, 2000.

14. Nicholas S. Downing et al., "Postmarket Safety Events Among Novel Therapeutics Approved by the US Food and Drug Administration Between 2001 and 2010," *Journal of the American Medical Association* 317 (2017): 1854–63.

15. Daniel C. Murrie et al., "Are Forensic Experts Biased by the Side That Retained Them?" *Psychological Science* 24 (2013): 1889–97.

16. Dan Fagin and Marianne Lavelle, *Toxic Deception* (Secaucus, NJ: Carol Publishing, 1996).

17. Richard A. Davidson, "Source of Funding and Outcome of Clinical Trials," *Journal of General Internal Medicine* 1 (May/June 1986): 155–58.

18. Lise L. Kjaergard and Bodil Als-Nielsen, "Association Between Competing Interests and Authors' Conclusions: Epidemiological Study of Randomised Clinical Trials Published In *BMJ,*" *British Medical Journal* 325 (August 3, 2002): 249–52. See also Krimsky, *Science in the Private Interest,* chapter 9, for a review of these and similar studies.

19. Alex Berenson et al., "Dangerous Data: Despite Warnings, Drug Giant Took Long Path to Vioxx Recall," *New York Times,* November 14, 2004.

20. Richard Horton, "The Lessons of MMR," *Lancet* 363 (2004): 747–49.

21. Andrew J. Wakefield, Peter Harvey, and John Linnell, "MMR—Responding to Retraction," *Lancet* 363 (2004): 1327–28.

22. Two of the best books on the subject are Paul Offit, *Deadly Choices: How the Anti-Vaccine Movement Threatens Us All* (New York: Basic Books, 2011), and Seth Mnookin, *The Panic Virus: The True Story Behind the Vaccine-Autism Controversy* (New York: Simon and Schuster, 2012). Thimerosal (variously spelled thimerosol and thimerserol) has been used commonly since the 1930s as a preservative in vaccines and many household products, such as cosmetics and eye drops. Anti-vaxxers maintain that the mercury contained in this preservative has toxic effects that cause autism and other diseases, but their arguments have largely been based on anecdotes, exaggerated fears, unsupported claims, and the antivaccine research conducted by Mark Geier and David Geier, president of a company specializing in litigating on behalf of alleged vaccine injury claimants. As for the research, in a study of all children born in Denmark between 1991 and 1998 (over half a million), the incidence of autism in vaccinated children was actually a bit lower than in unvaccinated children; see Kreesten M. Madsen et al., "A Population-Based Study of Measles, Mumps, and Rubella Vaccination and Autism," *New England Journal of Medicine* 347 (2002): 1477–82. Moreover, after vaccines containing thimerosal were removed from the market in

Denmark, there was no subsequent decrease in the incidence of autism; see Kreesten M. Madsen et al., "Thimerosal and the Occurrence of Autism: Negative Ecological Evidence from Danish Population-Based Data," *Pediatrics* 112 (2003): 604–6. See also L. Smeeth et al., "MMR Vaccination and Pervasive Developmental Disorders: A Case-Control Study," *Lancet* 364 (2004): 963–69. Another good review of the issues and studies is in Nick Paumgarten, "The Message of Measles," *New Yorker,* September 2, 2019.

23. Willem G. van Panhuis et al., "Contagious Diseases in the United States from 1888 to the Present," *New England Journal of Medicine* 369 (November 28, 2013): 2152–58; Paul A. Offit, *Do You Believe in Magic? The Sense and Nonsense of Alternative Medicine* (New York: HarperCollins, 2013), 139.

24. Brendan Nyhan et al., "Effective Messages in Vaccine Promotion: A Randomized Trial," *Pediatrics,* March 3, 2014, doi: 10.1542/peds.2013-2365. On their study of people who won't get flu shots on the mistaken belief that the vaccination can give them the flu, see Brendan Nyhan and Jason Reifler, "Does Correcting Myths About the Flu Vaccine Work? An Experimental Evaluation of the Effects of Corrective Information," *Vaccine* 33 (2015): 459–64.

25. http://www.prnewswire.com/news-releases/statement-from-dr-andrew-wakefield—no-fraud-no-hoax-no-profit-motive-113454389.html.

26. Clyde Haberman, "A Discredited Vaccine Study's Continuing Impact on Public Health," *New York Times,* February 1, 2015.

27. Dana and Loewenstein, "A Social Science Perspective on Gifts to Physicians from Industry."

28. Eric G. Campbell et al., "Physician Professionalism and Changes in Physician-Industry Relationships from 2004 to 2009," *Archives of Internal Medicine* 170 (November 8, 2010): 1820–26.

29. The Affordable Care Act mandated the Open Payments website, which went online October 1, 2014. Consumers can see how much money health-care professionals receive from drug companies as well as see whether their own physicians might have conflicts of interest.

See Charles Ornstein's reports at ProPublica, http://www.propublica. org/article/our-first-dive-into-the-new-open-paymentssystem?utm_ source=et&utm_medium=email&utm_campaign=dailynewsletter. His follow-up report on October 6, 2014, revealed that the Open Payments database of industry payments had underestimated the amount by about $1 billion.

30. Robert B. Cialdini, *Influence: The Psychology of Persuasion,* rev. ed. (New York: William Morrow, 1993).

31. Carl Elliott, "The Drug Pushers," *Atlantic Monthly,* April 2006, 82–93.

32. Carl Elliott, "Pharma Buys a Conscience," *American Prospect* 12 (September 24, 2001), www.prospect.org/print/V12/17/elliott-c.html.

33. C. Neil Macrae, Alan B. Milne, and Galen V. Bodenhausen, "Stereotypes as Energy-Saving Devices: A Peek Inside the Cognitive Toolbox," *Journal of Personality and Social Psychology* 66 (1994): 37–47.

34. Marilynn B. Brewer, "Social Identity, Distinctiveness, and In-Group Homogeneity," *Social Cognition* 11 (1993): 150–64.

35. Charles W. Perdue et al., "Us and Them: Social Categorization and the Process of Inter-Group Bias," *Journal of Personality and Social Psychology* 59 (1990): 475–86.

36. Henri Tajfel et al., "Social Categorization and Intergroup Behaviour," *European Journal of Social Psychology* 1 (1971): 149–78.

37. Nick Haslam et al., "More Human Than You: Attributing Humanness to Self and Others," *Journal of Personality and Social Psychology* 89 (2005): 937–50.

38. Gordon Allport, *The Nature of Prejudice* (Reading, MA: Addison-Wesley, 1979), 13–14.

39. Jeffrey W. Sherman et al., "Prejudice and Stereotype Maintenance Processes: Attention, Attribution, and Individuation," *Journal of Personality and Social Psychology* 89 (2005): 607–22.

40. Aaron Panofsky and Joan Donovan, "Genetic Ancestry Testing Among White Nationalists: From Identity Repair to Citizen Science," *Social Studies of Science*, July 2, 2019, https://doi. org/10.1177/0306312719861434.

41. Christian S. Crandall and Amy Eshelman, "A Justification-Suppression Model of the Expression and Experience of Prejudice," *Psychological Bulletin* 129 (2003): 425. See also Benoît Monin and Dale T. Miller, "Moral Credentials and the Expression of Prejudice," *Journal of Personality and Social Psychology* 81 (2001): 33–43. In their experiments, when people felt that their moral credentials as unprejudiced individuals were not in dispute — when they had been given a chance to disagree with blatantly sexist statements — they felt more justified in their subsequent vote to hire a man for a stereotypically male job.

42. For the interracial experiment, see Ronald W. Rogers and Steven Prentice-Dunn, "Deindividuation and Anger-Mediated Interracial Aggression: Unmasking Regressive Racism," *Journal of Personality and Social Psychology* 4 (1981): 63–73. For the English- and French-speaking Canadians, see James R. Meindl and Melvin J. Lerner, "Exacerbation of Extreme Responses to an Out-Group," *Journal of Personality and Social Psychology* 47 (1985): 71–84. On the studies of behavior toward Jews and gay men, see Steven Fein and Steven J. Spencer, "Prejudice as Self-Image Maintenance: Affirming the Self through Derogating Others," *Journal of Personality and Social Psychology* 73 (1997): 31–44.

43. Paul Jacobs, Saul Landau, and Eve Pell, *To Serve the Devil,* vol. 2, *Colonials and Sojourners* (New York: Vintage Books, 1971), 81.

44. Albert Speer, *Inside the Third Reich: Memoirs* (New York: Simon and Schuster, 1970), 291.

45. Doris Kearns Goodwin, *Team of Rivals: The Political Genius of Abraham Lincoln* (New York: Simon and Schuster, 2005).

46. Magruder, *An American Life,* 348.

3. MEMORY, THE SELF-JUSTIFYING HISTORIAN

1. George Plimpton, *Truman Capote* (New York: Doubleday, 1997), 306. We are taking Vidal's version of this story on the grounds that he has never had compunctions about talking about either subject — politics or bisexuality — and therefore had no motivation to distort the event in his memory.

2. Anthony G. Greenwald, "The Totalitarian Ego: Fabrication and Revision of Personal History," *American Psychologist* 35 (1980): 603–18.

3. Edward Jones and Rika Kohler, "The Effects of Plausibility on the Learning of Controversial Statements," *Journal of Abnormal and Social Psychology* 57 (1959): 315–20.

4. Michael Ross, "Relation of Implicit Theories to the Construction of Personal Histories," *Psychological Review* 96 (1989): 341–57; Anne E. Wilson and Michael Ross, "From Chump to Champ: People's Appraisals of Their Earlier and Present Selves," *Journal of Personality and Social Psychology* 80 (2001): 572–84; and Michael Ross and Anne E. Wilson, "Autobiographical Memory and Conceptions of Self: Getting Better All the Time," *Current Directions in Psychological Science* 12 (2003): 66–69.

5. E. S. Parker, L. Cahill, and J. L. McGaugh, "A Case of Unusual Autobiographical Remembering," *Neurocase* 12, no. 1 (February 2006): 35–49.

6. Marcia K. Johnson, Shahin Hashtroudi, and D. Stephen Lindsay, "Source Monitoring," *Psychological Bulletin* 114 (1993): 3–28; Karen J. Mitchell and Marcia K. Johnson, "Source Monitoring: Attributing Mental Experiences," in *The Oxford Handbook of Memory,* eds. E. Tulving and F.I.M. Craik (New York: Oxford University Press, 2000).

7. Mary McCarthy, *Memories of a Catholic Girlhood* (San Diego: Harcourt, Brace, 1957), 80–83.

8. Barbara Tversky and Elizabeth J. Marsh, "Biased Retellings of Events Yield Biased Memories," *Cognitive Psychology* 40 (2000): 1–38; see also Elizabeth J. Marsh and Barbara Tversky, "Spinning the Stories of Our Lives," *Applied Cognitive Psychology* 18 (2004): 491–503.

9. Brooke C. Feeney and Jude Cassidy, "Reconstructive Memory Related to Adolescent-Parent Conflict Interactions: The Influence of Attachment-Related Representations on Immediate Perceptions and Changes in Perceptions Over Time," *Journal of Personality and Social Psychology* 85 (2003): 945–55.

10. Daniel Offer et al., "The Altering of Reported Experiences," *Journal of*

the American Academy of Child and Adolescent Psychiatry 39 (2000) 735–
42. Several of the authors also wrote a book on this study. See Daniel
Offer, Marjorie Kaiz Offer, and Eric Ostrov, *Regular Guys: 34 Years
Beyond Adolescence* (New York: Kluwer Academic/Plenum, 2004).

11. On "mismemories" of sex, see Maryanne Garry et al., "Examining
Memory for Heterosexual College Students' Sexual Experiences Using
an Electronic Mail Diary," *Health Psychology* 21 (2002): 629–34. On
the overreporting of voting, see R. P. Abelson, Elizabeth D. Loftus, and
Anthony G. Greenwald, "Attempts to Improve the Accuracy of Self-
Reports of Voting," in *Questions About Questions: Inquiries into the
Cognitive Bases of Surveys,* ed. J. M. Tanur (New York: Russell Sage, 1992).
See also Robert F. Belli et al., "Reducing Vote Overreporting in Surveys:
Social Desirability, Memory Failure, and Source Monitoring," *Public
Opinion Quarterly* 63 (1999): 90–108. On misremembering donating
money, see Christopher D. B. Burt and Jennifer S. Popple, "Memorial
Distortions in Donation Data," *Journal of Social Psychology* 138 (1998):
724–33. College students' memories of their high-school grades are also
distorted in a positive direction; see Harry P. Bahrick, Lynda K. Hall,
and Stephanie A. Berger, "Accuracy and Distortion in Memory for High
School Grades," *Psychological Science* 7 (1996): 265–71.

12. J. Guillermo Villalobos, Deborah Davis, and Richard A. Leo, "His Story,
Her Story: Sexual Miscommunication, Motivated Remembering, and
Intoxication as Pathways to Honest False Testimony Regarding Sexual
Consent," in R. Burnett, ed., *Vilified: Wrongful Allegations of Sexual and
Child Abuse* (Oxford: Oxford University Press, 2016);
Deborah Davis and Elizabeth F. Loftus, "Remembering Disputed Sexual
Encounters," *Journal of Criminal Law and Criminology* 105 (2016): 811–
51.

13. Lisa K. Libby and Richard P. Eibach, "Looking Back in Time: Self-
Concept Change Affects Visual Perspective in Autobiographical
Memory," *Journal of Personality and Social Psychology* 82 (2002): 167–79.
See also Lisa K. Libby, Richard P. Eibach, and Thomas Gilovich, "Here's
Looking at Me: The Effect of Memory Perspective on Assessments of

Personal Change," *Journal of Personality and Social Psychology* 88 (2005): 50–62. The more consistent our memories are with our present selves, the more accessible they are. See Michael Ross, "Relation of Implicit Theories to the Construction of Personal Histories," *Psychological Review* 96 (1989): 341–57.

14. Michael Conway and Michael Ross, "Getting What You Want by Revising What You Had," *Journal of Personality and Social Psychology* 47 (1984): 738–48. Memory distortions take many different paths, but most are in the service of preserving our self-concepts and feelings about ourselves as good and competent people.

15. Anne E. Wilson and Michael Ross have shown how the self-justifying biases of memory help us move psychologically, in their words, from "chump to champ." We distance ourselves from our earlier "chumpier" incarnations if doing so allows us to feel better about how much we have grown, learned, and matured, but, like Haber, we feel close to earlier selves we thought were champs. Either way, we can't lose. See Wilson and Ross, "From Chump to Champ."

16. The full text of *Fragments,* along with the true story of Wilkomirski's life, is in Stefan Maechler, *The Wilkomirski Affair: A Study in Biographical Truth,* trans. John E. Woods (New York: Schocken, 2001). Maechler discusses the ways in which Wilkomirski drew on Kosinski's novel. For another investigation into Wilkomirski's life and the cultural issues of real and imagined memories, see Blake Eskin, *A Life in Pieces: The Making and Unmaking of Binjamin Wilkomirski* (New York: W. W. Norton, 2002).

17. The Will Andrews story is in Susan Clancy, *Abducted: How People Come to Believe They Were Kidnapped by Aliens* (Cambridge, MA: Harvard University Press, 2005). On the psychology of belief in alien abduction, see also Donald P. Spence, "Abduction Tales as Metaphors," *Psychological Inquiry* 7 (1996): 177–79. Spence interprets abduction memories as metaphors that have two powerful psychological functions: they encapsulate a set of free-floating concerns and anxieties that are widespread in today's political and cultural climate and have no ready

or easy remedy, and, by providing a shared identity for believers, they reduce the believers' feelings of alienation and powerlessness.

18. Maechler, *The Wilkomirski Affair,* 273.

19. Ibid., 27.

20. Ibid., 71. Wilkomirski accounted for having restless legs syndrome by telling a horrifying story: that when he was in Majdanek, he learned to keep his legs moving while he slept or otherwise "the rats would gnaw on them." But according to Tomasz Kranz, head of the research department at the Majdanek Museum, there were lice and fleas at the camp, but not rats (unlike other camps, such as Birkenau). See ibid., 169.

21. On the physical and psychological benefits of writing about previously undisclosed secrets and traumas, see James W. Pennebaker, *Opening Up* (New York: William Morrow, 1990).

22. On imagination inflation, see Elizabeth F. Loftus, "Memories of Things Unseen," *Current Directions in Psychological Science* 13 (2004): 145–47, and Loftus, "Imagining the Past," *Psychologist* 14 (2001): 584–87; Maryanne Garry et al., "Imagination Inflation: Imagining a Childhood Event Inflates Confidence That It Occurred," *Psychonomic Bulletin and Review* 3 (1996): 208–14; Giuliana Mazzoni and Amina Memon, "Imagination Can Create False Autobiographical Memories," *Psychological Science* 14 (2003): 186–88. On dreams, see Giuliana Mazzoni et al., "Changing Beliefs and Memories through Dream Interpretation," *Applied Cognitive Psychology* 2 (1999): 125–44.

23. Brian Gonsalves et al., "Neural Evidence that Vivid Imagining Can Lead to False Remembering," *Psychological Science* 15 (2004): 655–60. They found that the process of visually imagining a common object generates brain activity in regions of the cerebral cortex, which can lead to false memories of those imagined objects.

24. Mazzoni, "Imagination Can Create False Autobiographical Memories."

25. The effect is called "explanation inflation"; see Stefanie J. Sharman, Charles G. Manning, and Maryanne Garry, "Explain This: Explaining Childhood Events Inflates Confidence for Those Events," *Applied Cognitive Psychology* 19 (2005): 67–74. Preverbal children do the visual

equivalent of what adults do: They draw a picture of a completely implausible event, such as having a tea party in a hot-air balloon or swimming at the bottom of the ocean with a mermaid. After drawing these pictures, they often import them into their memories. A week later, they are far more likely than children who did not draw the pictures to say yes, that fanciful event really happened. See Deryn Strange, Maryanne Garry, and Rachel Sutherland, "Drawing Out Children's False Memories," *Applied Cognitive Psychology* 17 (2003): 607–19.

26. Maechler, *The Wilkomirski Affair,* 104.

27. Ibid., 100, 97 (emphasis ours).

28. Richard J. McNally, *Remembering Trauma* (Cambridge, MA: Harvard University Press, 2003), 233.

29. Michael Shermer, "Abducted!," *Scientific American* (February 2005): 33.

30. Clancy, *Abducted,* 51.

31. Ibid., 33, 34.

32. Giuliana Mazzoni and her colleagues showed in their laboratory how people can come to regard an impossible event (witnessing a demonic possession when they were children) as a plausible memory. One step in the process was reading about demonic possession in passages that said it was much more common than most people realized, accompanied by testimonials. See Giuliana Mazzoni, Elizabeth F. Loftus, and Irving Kirsch, "Changing Beliefs About Implausible Autobiographical Events: A Little Plausibility Goes a Long Way," *Journal of Experimental Psychology: Applied* 7 (2001): 51–59.

33. Clancy, *Abducted,* 143, 2.

34. Ibid., 50.

35. Richard McNally, personal communication with the authors.

36. Richard J. McNally et al., "Psychophysiologic Responding During Script-Driven Imagery in People Reporting Abduction by Space Aliens," *Psychological Science* 5 (2004): 493–97. See also Clancy, *Abducted,* and McNally, *Remembering Trauma,* for reviews of this and related research.

37. It is interesting that the autobiographies that once served as inspiring examples of a person's struggle to overcome racism, violence, disability,

exile, or poverty seem today so out of fashion. Modern memoirists strive to outdo one another in the gruesome details of their lives. For an eloquent essay on this theme, see Francine Prose, "Outrageous Misfortune," her review of Jeannette Walls's *The Glass Castle: A Memoir* for the *New York Times Book Review,* March 13, 2005. Prose begins, "Memoirs are our modern fairy tales, the harrowing fables of the Brothers Grimm reimagined from the perspective of the plucky child who has, against all odds, evaded the fate of being chopped up, cooked and served to the family for dinner."

38. Ellen Bass and Laura Davis, *The Courage to Heal: A Guide for Women Survivors of Child Sexual Abuse* (New York: Harper and Row, 1988), 173.

39. For the best full account of this story, see Moira Johnston, *Spectral Evidence: The Ramona Case: Incest, Memory, and Truth on Trial in Napa Valley* (Boston: Houghton Mifflin, 1997).

40. Mary Karr, "His So-Called Life," *New York Times,* January 15, 2006.

4. GOOD INTENTIONS, BAD SCIENCE: THE CLOSED LOOP OF CLINICAL JUDGMENT

1. The story of Grace was told to us by psychologist Joseph de Rivera, who interviewed her and others in his research on the psychology of recanters. See Joseph de Rivera, "The Construction of False Memory Syndrome: The Experience of Retractors," *Psychological Inquiry* 8 (1997): 271–92; and Joseph de Rivera, "Understanding Persons Who Repudiate Memories Recovered in Therapy," *Professional Psychology: Research and Practice* 31 (2000): 378–86.

2. The most comprehensive history of the recovered-memory epidemic remains Mark Pendergrast, *Victims of Memory,* 2nd ed. (Hinesburg, VT: Upper Access Press, 1996; revised and expanded for a HarperCollins British edition, 1996). See also Richard J. Ofshe and Ethan Watters, *Making Monsters: False Memory, Psychotherapy, and Sexual Hysteria* (New York: Scribner's, 1994); Elizabeth Loftus and Katherine Ketcham, *The Myth of Repressed Memory* (New York: St. Martin's Press, 1994); and Frederick Crews, ed., *Unauthorized Freud: Doubters Confront a Legend*

(New York: Viking, 1998). For an excellent sociology of hysterical epidemics and moral panics, see Philip Jenkins, *Intimate Enemies: Moral Panics in Contemporary Great Britain* (Hawthorne, NY: Aldine de Gruyter, 1992).

The woman who claimed that her father molested her from the ages of five to twenty-three, Laura B., sued her father, Joel Hungerford, in the state of New Hampshire in 1995. She lost.

3. For analyses of the rise and fall of MPD, see also Joan Acocella, *Creating Hysteria: Women and Multiple Personality Disorder* (San Francisco: Jossey-Bass, 1999). On hypnosis and other means of creating false memories of abduction, multiple personality disorder, and child abuse, see Nicholas P. Spanos, *Multiple Identities and False Memories: A Sociocognitive Perspective* (Washington, DC: American Psychological Association, 1996).

4. Judith Levine, "Bernard Baran, RIP," *Seven Days,* September 13, 2014. As a gay man and convicted child molester in prison, Baran suffered twenty-one years of violence before he was freed on retrial. In 2014, eight years after his release, he died of an aneurysm.

5. Three of the best books on the daycare scandals and claims of widespread cults that were promoting ritual satanic sexual abuse are Debbie Nathan and Michael Snedeker, *Satan's Silence: Ritual Abuse and the Making of a Modern American Witch Hunt* (New York: Basic Books, 1995); Stephen J. Ceci and Maggie Bruck, *Jeopardy in the Courtroom: A Scientific Analysis of Children's Testimony* (Washington, DC: American Psychological Association, 1995); and, a superb and detailed account of the McMartin case, Richard Beck, *We Believe the Children: A Moral Panic in the 1980s* (New York: Public Affairs, 2015). Dorothy Rabinowitz, a *Wall Street Journal* editorial writer, was the first to publicly question the conviction of Kelly Michaels and get her case reopened; see Rabinowitz, *No Crueler Tyrannies: Accusation, False Witness, and Other Terrors of Our Times* (New York: Free Press, 2003).

6. In 2005, a Boston jury convicted a seventy-four-year-old former priest, Paul Shanley, of sexually molesting twenty-seven-year-old Paul Busa

when Busa was six. This claim followed upon the Church scandals that had revealed hundreds of documented cases of pedophile priests, so emotions understandably ran high against the priests and the Church's policy of covering up the accusations. Yet the *sole* evidence in Shanley's case was Busa's memories, which, Busa said, he recovered in vivid flashbacks after reading a *Boston Globe* article on Shanley. There was no corroborating evidence presented at the trial and indeed much that disputed Busa's claims. See Jonathan Rauch, "Is Paul Shanley Guilty? If Paul Shanley Is a Monster, the State Didn't Prove It," *National Journal,* March 12, 2005, and JoAnn Wypijewski, "The Passion of Father Paul Shanley," *Legal Affairs* (September/October 2004). Other skeptical reporters included Daniel Lyons of *Forbes,* Robin Washington of the *Boston Herald,* and Michael Miner of the *Chicago Reader.* For an even more sensational story of the conviction of a man based almost entirely on repressed-and-recovered memories, consider the case of Jerry Sandusky: Mark Pendergrast, *The Most Hated Man in America: Jerry Sandusky and the Rush to Judgment* (Mechanicsburg, PA: Sunbury Press, 2017). For a summary of the case, see Frederick Crews, "Trial by Therapy: The Jerry Sandusky Case Revisited," *Skeptic,* https://www.skeptic.com/reading_room/trial-by-therapy-jerry-sandusky-case-revisited/.

7. Debbie Nathan, *Sybil Exposed: The Extraordinary Story Behind the Famous Multiple Personality Case* (New York: Free Press, 2011).

8. Some studies find that combined approaches — medication plus cognitive-behavior therapy (CBT) — are most effective; others find that CBT alone does as well. For a review of the issues and bibliography of research studies, see the American Psychological Association Presidential Task Force on Evidence-Based Practice, "Evidence-Based Practice in Psychology," *American Psychologist* 61 (2006): 271–83. See also Dianne Chambless et al., "Update on Empirically Validated Therapies," *Clinical Psychologist* 51 (1998): 3–16, and Steven D. Hollon, Michael E. Thase, and John C. Markowitz, "Treatment and Prevention of Depression," *Psychological Science in the Public Interest* 3 (2002):

39–77. These articles contain excellent references regarding empirically validated forms of psychotherapy for different problems.

9. Tanya M. Luhrmann, *Of Two Minds: The Growing Disorder in American Psychiatry* (New York: Knopf, 2000). Her findings echo precisely what Jonas Robitscher described about his profession twenty years earlier in *The Powers of Psychiatry* (Boston: Houghton Mifflin, 1980).

10. For an excellent review of the issues and the rise of pseudoscientific methods and practices in psychotherapy — including unvalidated assessment tests, treatments for autism and ADHD, and popular therapies — see Scott O. Lilienfeld, Steven Jay Lynn, and Jeffrey M. Lohr, eds., *Science and Pseudoscience in Clinical Psychology,* 2nd ed. (New York: Guilford, 2015). And for the other side of the story, articles on the most important contributions of clinical science, see Scott O. Lilienfeld and William T. O'Donohue, eds., *The Great Ideas of Clinical Science* (New York: Routledge, 2007).

11. On evidence that hypnosis is effective for a large number of acute and chronic pain conditions, see David R. Patterson and Mark P. Jensen, "Hypnosis and Clinical Pain," *Psychological Bulletin* 29 (2003): 495–521. Hypnosis can also add to the effectiveness of cognitive-behavioral techniques for losing weight, quitting smoking, and other behavior problems; see Irving Kirsch, Guy Montgomery, and Guy Sapirstein, "Hypnosis as an Adjunct to Cognitive-Behavioral Psychotherapy: A Meta-Analysis," *Journal of Consulting and Clinical Psychology* 2 (1995): 214–20. But the evidence is overwhelming that hypnosis is unreliable as a way of retrieving memories, which is why the American Psychological Association and the American Medical Association oppose the use of "hypnotically refreshed" testimony in courts of law. See Steven Jay Lynn et al., "Constructing the Past: Problematic Memory Recovery Techniques in Psychotherapy," in Lilienfeld, Lynn, and Lohr, *Science and Pseudoscience in Clinical Psychology;* and John F. Kihlstrom, "Hypnosis, Delayed Recall, and the Principles of Memory," *International Journal of Experimental Hypnosis* 42 (1994): 337–45.

12. Paul Meehl, "Psychology: Does Our Heterogeneous Subject Matter Have Any Unity?," *Minnesota Psychologist* (Summer 1986): 4.

13. Bessel van der Kolk's deposition was taken by attorney and psychologist R. Christopher Barden in van der Kolk's office in Boston, Massachusetts, December 27 and 28, 1996. Barden has posted the deposition online; see "Full Text of 'Bessel van der Kolk, Scientific Dishonesty, and the Mysterious Disappearing Coauthor,'" https://archive.org/stream/BesselVanDerKolkScientificDishonestyTheMysteriousDisappearing/VanDerKolk_djvu.txt.

14. John F. Kihlstrom, "An Unbalanced Balancing Act: Blocked, Recovered, and False Memories in the Laboratory and Clinic," *Clinical Psychology: Science and Practice* 11 (2004). He added that "if confidence were an adequate criterion for validity, Binjamin Wilkomirski might have gotten a Pulitzer Prize for history."

15. Dr. Courtois's testimony was given on November 14, 2014, in the case of *John Doe v. Society of Missionaries of the Sacred Heart,* Chicago, Illinois.

16. See Deena S. Weisberg et al., "The Seductive Allure of Neuroscience Explanations," *Journal of Cognitive Neuroscience* 20 (2008): 470–77.

17. Sigmund Freud, "The Dissolution of the Oedipus Complex," in *The Standard Edition of the Complete Psychological Works of Sigmund Freud,* ed. J. Strachey, vol. 19 (London: Hogarth, 1924).

18. Rosenzweig wrote: "On two separate occasions (1934 and 1937), first in gothic script and then in English, Freud made a similar negative response to any attempts to explore psychoanalytic theory by laboratory methods. This exchange clearly underscored Freud's distrust of, if not opposition to, experimental approaches to the validation of his clinically derived concepts. Freud consistently believed that the clinical validation of his theories, which were based originally and continuously on his self-analysis, left little to be desired from other sources of support." In Saul Rosenzweig, "Letters by Freud on Experimental Psychodynamics," *American Psychologist* 52 (1997): 571. See also Saul Rosenzweig, "Freud and Experimental Psychology: The Emergence of Idio-Dynamics," in

A Century of Psychology as Science, eds. S. Koch and D. E. Leary (New York: McGraw-Hill, 1985). This book was reissued by the American Psychological Association in 1992.

19. Lynn et al., "Constructing the Past."

20. Michael Nash offers one example in his article "Memory Distortion and Sexual Trauma: The Problem of False Negatives and False Positives," *International Journal of Clinical and Experimental Hypnosis* 42 (1994): 346–62.

21. McNally, *Remembering Trauma,* 275.

22. The recovered-memory advocates in question are Daniel Brown, Alan W. Scheflin, and D. Corydon Hammond, authors of *Memory, Trauma Treatment, and the Law* (New York: W. W. Norton, 1998); their rendering of the Camp Erika study is on page 156. For a review of this book that documents its authors' long association with the recovered-memory movement, their belief in the prevalence of satanic ritual-abuse cults, and their endorsement of the use of hypnosis to "recover" memories of abuse and generate multiple personalities, see Frederick Crews's "The Trauma Trap," *New York Review of Books* 51 (March 11, 2004). This essay has been reprinted, with other writings exposing the fallacies of the recovered-memory movement, in Frederick Crews, *Follies of the Wise* (Emeryville, CA: Shoemaker and Hoard, 2006).

23. Rosemary Basson et al., "Efficacy and Safety of Sildenafil Citrate in Women with Sexual Dysfunction Associated with Female Sexual Arousal Disorder," *Journal of Women's Health and Gender-Based Medicine* 11 (May 2002): 367–77.

24. Joan Kaufman and Edward Zigler, "Do Abused Children Become Abusive Parents?" *American Journal of Orthopsychiatry* 57 (1987): 186–92. Ever since Freud, there has been a widespread cultural assumption that childhood trauma always, inevitably, produces adult psychopathology. Research has shattered this assumption too. Psychologist Ann Masten has observed that most people assume there is something special and rare about the children who recover from adversity. But "the great surprise" of the research, she concluded, is how

ordinary resilience is. Most children are remarkably resilient, eventually overcoming even the effects of war, childhood illness, having abusive or alcoholic parents, early deprivation, or being sexually molested. See Ann Masten, "Ordinary Magic: Resilience Processes in Development," *American Psychologist* 56 (2001): 227–38.

25. William Friedrich et al., "Normative Sexual Behavior in Children: A Contemporary Sample," *Pediatrics* 101 (1988): 1–8. See also www.pediatrics.org/cgi/content/full/101/4/e9. For an excellent review of the behavioral-genetics research on the stability of temperament regardless of a child's experiences, see Judith Rich Harris, *The Nurture Assumption* (New York: Free Press, 1998). That nonabused children often have nightmares and other symptoms of anxiety, see McNally, *Remembering Trauma.*

26. Kathleen A. Kendall-Tackett, Linda M. Williams, and David Finkelhor, "Impact of Sexual Abuse on Children: A Review and Synthesis of Recent Empirical Studies," *Psychological Bulletin* 113 (1992): 164–80. The researchers also found, not surprisingly, that the children's symptoms were related to the severity, duration, and frequency of the abuse, whether force had been used, the perpetrator's relationship to the child, and the degree of support from the mother. In contrast to the predictions of recovered-memory therapists, about two-thirds of the victimized children recovered during the first twelve to eighteen months.

27. In reviewing the research, Glenn Wolfner, David Faust, and Robyn Dawes concluded, "There is simply no scientific evidence available that would justify clinical or forensic diagnosis of abuse on the basis of doll play"; see their paper "The Use of Anatomically Detailed Dolls in Sexual Abuse Evaluations: The State of the Science," *Applied and Preventive Psychology* 2 (1993): 1–11.

28. When the little girl was asked if this really happened, she said, "Yes, it did." When her father and the experimenter both tried to reassure her by saying, "Your doctor doesn't do those things to little girls. You were just fooling. We know he didn't do those things," the child clung tenaciously to her claims. "Thus, repeated exposure to the doll, with

minimal suggestions," the researchers cautioned, "resulted in highly sexualized play for this one child." Maggie Bruck et al., "Anatomically Detailed Dolls Do Not Facilitate Preschoolers' Reports of a Pediatric Examination Involving Genital Touching," *Journal of Experimental Psychology: Applied* 1 (1995): 95–109.

29. Thomas M. Horner, Melvin J. Guyer, and Neil M. Kalter, "Clinical Expertise and the Assessment of Child Sexual Abuse," *Journal of the American Academy of Child and Adolescent Psychiatry* 32 (1993): 925–31; Thomas M. Horner, Melvin J. Guyer, and Neil M. Kalter, "The Biases of Child Sexual Abuse Experts: Believing Is Seeing," *Bulletin of the American Academy of Psychiatry and the Law* 21 (1993): 281–92.

30. Many decades ago, Paul Meehl showed that relatively simple mathematical formulas outperformed clinicians' intuitive judgments in predicting patients' outcomes; see Paul E. Meehl, *Clinical versus Statistical Prediction: A Theoretical Analysis and a Review of the Evidence* (Minneapolis: University of Minnesota Press, 1954); and Robyn Dawes, David Faust, and Paul E. Meehl, "Clinical versus Actuarial Judgment," *Science* 243 (1989): 1668–74. Meehl's findings have been repeatedly reconfirmed. See Howard Grob, *Studying the Clinician: Judgment Research and Psychological Assessment* (Washington, DC: American Psychological Association, 1998).

31. Our account of the Kelly Michaels case is based largely on Ceci and Bruck, *Jeopardy in the Courtroom,* and Pendergrast, *Victims of Memory.* See also Maggie Bruck and Stephen Ceci, "Amicus Brief for the Case of *State of New Jersey v. Margaret Kelly Michaels,* Presented by Committee of Concerned Social Scientists," *Psychology, Public Policy, and Law* 1 (1995).

32. Pendergrast, *Victims of Memory,* 423.

33. Jason J. Dickinson, Debra A. Poole, and R. L. Laimon, "Children's Recall and Testimony," in *Psychology and Law: An Empirical Perspective,* eds. N. Brewer and K. Williams (New York: Guilford, 2005). See also Debra A. Poole and D. Stephen Lindsay, "Interviewing Preschoolers: Effects of Nonsuggestive Techniques, Parental Coaching, and

Leading Questions on Reports of Nonexperienced Events," *Journal of Experimental Child Psychology* 60 (1995): 129–54.

34. Sena Garven et al., "More Than Suggestion: The Effect of Interviewing Techniques from the McMartin Preschool Case," *Journal of Applied Psychology* 83 (1998): 347–59; and Sena Garven, James M. Wood, and Roy S. Malpass, "Allegations of Wrongdoing: The Effects of Reinforcement on Children's Mundane and Fantastic Claims," *Journal of Applied Psychology* 85 (2000): 38–49.

35. Gabrielle F. Principe et al., "Believing Is Seeing: How Rumors Can Engender False Memories in Preschoolers," *Psychological Science* 17 (2006): 243–48.

36. Debbie Nathan, "I'm Sorry," *Los Angeles Times Magazine,* October 30, 2005.

37. Debra A. Poole and Michael E. Lamb, *Investigative Interviews of Children* (Washington, DC: American Psychological Association, 1998). Their work became the basis of new protocols drafted by the State of Michigan Governor's Task Force on Children's Justice and Family Independence Agency and the National Institute of Child Health and Human Development (NICHD), which prepared an investigative interview protocol that is widely used in research and assessment: https://youth.gov/content/nichd-investigative-interview-protocol. See Michael E. Lamb, Yael Orbach, Irit Hershkowitz, Phillip W. Esplin, and Dvora Horowitz, "Structured Forensic Interview Protocols Improve the Quality and Informativeness of Investigative Interviews with Children: A Review of Research Using the NICHD Investigative Interview Protocol," *Child Abuse and Neglect* 31 (2007): 1201–31.

38. Ellen Bass and Laura Davis, *The Courage to Heal: A Guide for Women Survivors of Child Sexual Abuse* (New York: Harper and Row, 1998), 18.

39. In a study conducted in the mid-1990s, researchers drew random samples of American clinical psychologists with PhDs from names listed in the National Register of Health Service Providers in Psychology. They asked respondents how often they regularly used certain techniques specifically "to help clients recover memories of

sexual abuse": hypnosis, age regression, dream interpretation, guided imagery related to abuse situations, and interpreting physical symptoms as evidence of abuse. Slightly more than 40 percent said they used dream interpretation; about 30 percent said they used hypnosis; the fewest, but still about 20 percent, used age regression. About the same percentages disapproved of using these techniques; those in the middle apparently had no opinion. See Debra A. Poole et al., "Psychotherapy and the Recovery of Memories of Childhood Sexual Abuse: U.S. and British Practitioners' Opinions, Practices, and Experiences," *Journal of Consulting and Clinical Psychology* 63 (1995): 426–37. Yet the scientist-practitioner gap continues; see Lawrence Patihis et al., "Are the 'Memory Wars' Over? A Scientist-Practitioner Gap in Beliefs About Repressed Memory," *Psychological Science* 25 (2014): 519–30.

40. According to a meta-analysis of the leading studies, the notion that childhood sexual abuse is a leading cause of eating disorders has not been supported by empirical evidence. See Eric Stice, "Risk and Maintenance Factors for Eating Pathology: A Meta-Analytic Review," *Psychological Bulletin* 128 (2002): 825–48.

41. Patihis et al., "Are the 'Memory Wars' Over?"

42. Henry Otgaar, Mark L. Howe, Lawrence Patihis, Harald Merckelbach, Steven Jay Lynn, Scott O. Lilienfeld, and Elizabeth F. Loftus, "The Return of the Repressed: The Persistent and Problematic Claims of Long-Forgotten Trauma," *Perspectives on Psychological Science* 14, no. 6 (2019): 1072–95.

43. Some have simply shifted focus. Bessel van der Kolk lost his affiliation with Harvard Medical School, and his lab at Massachusetts General Hospital was closed, but he still believes that repressed memories are a common feature of traumatic stress disorders. He went on to bypass mental mechanisms as explanations and argue that traumatic memories get "stuck in the machine" and are expressed in the body, which "betrayed" the sufferer during the traumatic episode. See Jeneen Interlandi, "How Do You Heal a Traumatized Mind?," *New York Times Magazine,* May 25, 2014.

44. Richard J. McNally, "Troubles in Traumatology," *Canadian Journal of Psychiatry* 50 (2005): 815.

45. John Briere made this statement at the Twelfth International Congress on Child Abuse and Neglect in 1998, in Auckland, New Zealand. These remarks were reported by the *New Zealand Herald,* September 9, 1998. The paper quoted Briere as saying that "missing memories of abuse are reasonably common, but evidence suggests that false memories of abuse are quite uncommon." See http://www.menz.org.nz/ Casualties/1998%20newsletters/Oct%2098.htm.

46. Pendergrast, *Victims of Memory,* 567.

47. Hammond made these remarks in his presentation "Investigating False Memory for the Unmemorable: A Critique of Experimental Hypnosis and Memory Research," at the Fourteenth International Congress of Hypnosis and Psychosomatic Medicine, San Diego, June 1997.

48. One group of psychiatrists and other clinical experts asked the United States Department of Justice to pass a law making it illegal to publish excerpts of children's testimony in the actual daycare cases. The DOJ refused. Basic Books was threatened with an injunction if it published Debbie Nathan and Michael Snedeker's *Satan's Silence,* an exposé of the daycare hysteria; Basic Books did not comply with the threateners' demands. The American Psychological Association was threatened with a lawsuit if it published Stephen Ceci and Maggie Bruck's *Jeopardy in the Courtroom;* the APA delayed publication for several months. (Our source is personal communications from the investigators involved.)

49. But how do you kill the messenger when there are hundreds of messengers? One way to resolve the dissonance between "I'm certain I'm right" and "I'm in a small minority" is to claim that scientific consensus reflects a "conspiracy" to suppress the truth of child sexual abuse. For example, political scientist Ross Cheit claims that a conspiracy of journalists, defense attorneys, social scientists, and critics of the criminal justice system invented a "witch-hunt narrative." He maintains that there was no witch-hunt against those hundreds of daycare workers; most of those who were convicted and later released were, he maintains,

guilty. But Cheit cherry-picked the evidence, looking for arguments to support his allegations and distorting or omitting the evidence he didn't like. See Ross E. Cheit, *The Witch-Hunt Narrative* (New York: Oxford University Press, 2014). For readers interested in detailed rebuttals of this book, see "The Witch Hunt Narrative: Rebuttal," http://www.ncrj.org/resources-2/response-to-ross-cheit/the-witch-hunt-narrative-rebuttal/; and Cathy Young, "The Return of Moral Panic," *Reason,* October 25, 2014, http://reason.com/archives/2014/10/25/the-return-of-moral-panic.

50. To our knowledge, neither Bass nor Davis has ever acknowledged that she was wrong in any of the basic claims about memory and trauma; they have never admitted that their ignorance of psychological science might have caused them to overreach. In the preface to the third edition of *The Courage to Heal,* Bass and Davis responded to the scientific criticism directed at their book and attempted to justify their claims of expertise despite their lack of professional training: "As authors, we have been criticized for our lack of academic credentials. But you do not have to have a Ph.D. to listen carefully and compassionately to another human being." That is true, but training in science might prevent all those compassionate listeners from leaping to unwarranted, implausible, potentially harmful conclusions. In the book's twentieth-anniversary edition, published in 2008, the section "Honoring the Truth" has been removed. Not because it was wrong, the authors hastened to explain, but "to make room for new stories and information about healing," including a new boxed feature called "The Essential Truth of Memory."

51. National Public Radio, *This American Life,* episode 215, June 16, 2002.

5. LAW AND DISORDER

1. Timothy Sullivan, *Unequal Verdicts: The Central Park Jogger Trials* (New York: Simon and Schuster, 1992). See also Sarah Burns's *The Central Park Five: A Chronicle of a City Wilding* (New York: Knopf, 2012). A film based on her book, directed by Ken Burns, Sarah Burns,

and David McMahon, was released in 2012, and Ava DuVernay's dramatized film *When They See Us* came out in 2019.

2. Reyes confessed because, entirely by chance, he met one of the convicted defendants, Kharey Wise, in prison and apparently came to feel guilty about Wise's wrongful incarceration. Later he told prison officials that he had committed a crime for which others had been wrongly convicted, and a reinvestigation began. Steven A. Drizin and Richard A. Leo, "The Problem of False Confessions in the Post-DNA World," *North Carolina Law Review* 82 (2004): 891–1008.

3. Stuart Jeffries, "The Rapist Hunter," *Guardian,* February 26, 2004.

4. Linda Fairstein, "Netflix's False Story of the Central Park Five," *Wall Street Journal,* June 10, 2019.

5. See www.innocenceproject.org for latest updates and the classic book by Barry Scheck, Peter Neufeld, and Jim Dwyer, *Actual Innocence* (New York: Doubleday, 2000).

6. Samuel R. Gross, "How Many False Convictions Are There? How Many Exonerations Are There?," University of Michigan Public Law Research Paper No. 316, February 26, 2013, available at https://papers.ssrn.com/sol3/papers.cfm?abstract_id=2225420. See C. R. Huff and M. Killias, eds., *Wrongful Convictions and Miscarriages of Justice: Causes and Remedies in North American and European Criminal Justice Systems* (New York: Routledge, 2013).

7. See http://www.law.umich.edu/special/exoneration/Pages/about.aspx. As DNA has become more widely used in forensic investigations before trial, some legal scholars have predicted that the number of clear, DNA-based exonerations will fade, and the focus will turn to other bases for overturning wrongful convictions. For a thoughtful assessment of the difference between "factual innocence" and "exoneration" and how this might apply to the innocence movement, see Richard A. Leo, "Has the Innocence Movement Become an Exoneration Movement? The Risks and Rewards of Redefining Innocence," in Daniel Medwed, ed., *Wrongful Convictions and the DNA Revolution: Twenty-Five Years of*

Freeing the Innocent (Cambridge, MA: Cambridge University Press, 2017), 57–83.

8. Quoted in Richard Jerome, "Suspect Confessions," *New York Times Magazine,* August 13, 1995.

9. Daniel S. Medwed, "The Zeal Deal: Prosecutorial Resistance to Post-Conviction Claims of Innocence," *Boston University Law Review* 84 (2004): 125. Medwed analyzes the institutional culture of many prosecutors' offices that makes it difficult for prosecutors to admit mistakes and correct them.

10. Joshua Marquis, "The Innocent and the Shammed," *New York Times,* January 26, 2006. As of 2014, he was unmoved by evidence of wrongful convictions. If Samuel Gross's estimates that 4.1 percent of all death row defendants were falsely convicted, Marquis told an interviewer, "I would quit my job and become a Buddhist monk if it was one-fifth accurate." But according to Gross, "one fifth accurate" *is* agreed as the lowest estimate. See https://www.nytimes.com/2014/05/02/science/convictions-of-4-1-percent-facing-death-said-to-be-false.html.

11. Registry of Prosecutorial Misconduct, www.prosecutorintegrity.org/registry/. See Kathleen M. Ridolfi and Maurice Possley, "Preventable Error: A Report on Prosecutorial Misconduct in California 1997–2009," a VERITAS Initiative Report by the Northern California Innocence Project, 2010.

12. *Harmful Error: Investigating America's Local Prosecutors,* published by the Center for Public Integrity, Summer 2003, reports on their analysis of 11,452 cases across the nation in which appellate court judges reviewed charges of prosecutorial misconduct.

13. Quoted in Mike Miner, "Why Can't They Admit They Were Wrong?," *Chicago Reader,* August 1, 2003.

14. The main problem with the voice stress analyzer is that the confirmation bias gets in the way. If you think the suspect is guilty, you interpret the microtremors as signs of lying, and if you think the suspect is innocent, you pay them no attention. One major study, "The Validity and Comparative Accuracy of Voice Stress Analysis," found (contrary to

the title) that "the CVSA examiners were not able to distinguish truth tellers from deceivers at higher than chance levels." See https://www.polygraph.org/assets/docs/VoiceStressStudies/palmatier%20study.pdf.

15. Paul E. Tracy, *Who Killed Stephanie Crowe?* (Dallas, TX: Brown Books, 2003), 334.

16. The account of Vic Caloca's involvement in the case, including the quotes by him, comes from a story written by investigative reporters John Wilkens and Mark Sauer, "A Badge of Courage: In the Crowe Case, This Cop Ignored the Politics While Pursuing Justice," *San Diego Union-Tribune,* July 11, 2004. Druliner's quote is in Mark Sauer and John Wilkens, "Tuite Found Guilty of Manslaughter," *San Diego Union-Tribune,* May 27, 2004.

17. Deanna Kuhn, Michael Weinstock, and Robin Flaton, "How Well Do Jurors Reason? Competence Dimensions of Individual Variation in a Juror Reasoning Task," *Psychological Science* 5 (1994): 289–96.

18. Don DeNevi and John H. Campbell, *Into the Minds of Madmen: How the FBI's Behavioral Science Unit Revolutionized Crime Investigation* (Amherst, NY: Prometheus Books, 2004), 33. This book is, unintentionally, a case study of the unscientific training of the FBI's Behavioral Science Unit.

19. Tracy, *Who Killed Stephanie Crowe?,* 184.

20. Ralph M. Lacer, interview by Connie Chung, *Eye to Eye with Connie Chung,* CBS, broadcast January 13, 1994.

21. Introductory comments by Steven Drizin, "Prosecutors Won't Oppose Tankleff's Hearing," *New York Times,* May 13, 2004.

22. Edward Humes, *Mean Justice* (New York: Pocket Books, 1999), 181.

23. Andrew J. McClurg, "Good Cop, Bad Cop: Using Cognitive Dissonance Theory to Reduce Police Lying," *U.C. Davis Law Review* 32 (1999): 395, 429.

24. This excuse is so common that it, too, has spawned a new term: *dropsy testimony.* David Heilbroner, a former New York assistant district attorney, wrote: "In dropsy cases, officers justify a search by the oldest of means: they lie about the facts: "As I was coming around the corner I

saw the defendant drop the drugs on the sidewalk, so I arrested him." It was an old line known to everyone in the justice system. One renowned federal judge many years ago complained that he had read the same testimony in too many cases for it to be believed any longer as a matter of law." David Heilbroner, *Rough Justice: Days and Nights of a Young D.A.* (New York: Pantheon, 1990), 29.

25. McClurg, "Good Cop, Bad Cop," 391.

26. Norm Stamper, *Breaking Rank: A Top Cop's Exposé of the Dark Side of American Policing* (New York: Nation Books, 2005), and Norm Stamper, "Let Those Dopers Be," *Los Angeles Times,* October 16, 2005.

27. McClurg, "Good Cop, Bad Cop," 413, 415.

28. In Suffolk County, New York, in September 1988, homicide detective K. James McCready was summoned to a home where he found the body of Arlene Tankleff, who had been stabbed and beaten to death, and her unconscious husband, Seymour, who had also had been brutally attacked. (He died a few weeks later.) Within hours, McCready declared that he had solved the case: the killer was the couple's son, Martin, age seventeen. During the interrogation, McCready repeatedly told Martin that he knew he had killed his parents because his father had briefly regained consciousness and told police that Marty was his attacker. This was a lie. "I used trickery and deceit," McCready said. "I don't think he did it. I know he did it." The teenager finally confessed that he must have killed his parents while in a blackout. When the family lawyer arrived at the police station, Martin Tankleff immediately disavowed the confession and never signed it, but it was enough to convict him. Martin was sentenced to fifty years to life in prison. Bruce Lambert, "Convicted of Killing His Parents, but Calling a Detective the Real Bad Guy," *New York Times,* April 4, 2004. Tankleff was released from prison in 2007 after his conviction was overturned. In 2014 he was awarded $3.4 million from the state under a settlement of his wrongful-conviction lawsuit. His civil rights lawsuit against McCready and Suffolk County is pending.

29. Tracy, *Who Killed Stephanie Crowe?,* 175.

30. Fred E. Inbau et al., *Criminal Interrogation and Confessions,* 5th ed. (Burlington, MA: Jones and Bartlett Learning, 2011), xi.
31. Ibid., 352.
32. Ibid., 5.
33. One of the most thorough dissections of the Reid Technique and the Inbau et al. manual is Deborah Davis and William T. O'Donohue, "The Road to Perdition: 'Extreme Influence' Tactics in the Interrogation Room," in *Handbook of Forensic Psychology,* eds. W. T. O'Donohue and E. Levensky (New York: Elsevier Academic Press, 2004), 897–996. See also Timothy E. Moore and C. Lindsay Fitzsimmons, "Justice Imperiled: False Confessions and the Reid Technique," *Criminal Law Quarterly* 57 (2011): 509–42; Lesley King and Brent Snook, "Peering Inside a Canadian Interrogation Room: An Examination of the Reid Model of Interrogation, Influence Tactics, and Coercive Strategies," *Criminal Justice and Behavior* 36 (2009): 674–94.
34. Louis C. Senese, *Anatomy of Interrogation Themes: The Reid Technique of Interviewing* (Chicago: John E. Reid and Associates, 2005), 32.
35. Saul Kassin, "On the Psychology of Confessions: Does Innocence Put Innocents at Risk?," *American Psychologist* 60 (2005): 215–28.
36. Saul M. Kassin and Christina T. Fong, "I'm Innocent! Effects of Training on Judgments of Truth and Deception in the Interrogation Room," *Law and Human Behavior* 23 (1999): 499–516. In another study, Kassin and his colleagues recruited prison inmates who were instructed to give a full confession of their own crime and a made-up confession of a crime committed by another inmate. College students and police investigators judged the videotaped confessions. The overall accuracy rate did not exceed chance, but the police were more confident in their judgments. See Saul M. Kassin, Christian A. Meissner, and Rebecca J. Norwick, "'I'd Know a False Confession If I Saw One': A Comparative Study of College Students and Police Investigators," *Law and Human Behavior* 29 (2005): 211–27.
37. This is why innocent people are more likely than guilty people to waive their Miranda rights to silence and to having a lawyer. In one of

Saul Kassin's experiments, seventy-two participants who were guilty
or innocent of a mock theft of a hundred dollars were interrogated by
a male detective whose demeanor was neutral, sympathetic, or hostile
and who tried to get them to give up their Miranda rights. Those
who were innocent were far more likely to sign a waiver than those
who were guilty, and by a large margin — 81 percent to 36 percent.
Two-thirds of the innocent suspects even signed the waiver when the
detective adopted a hostile pose, shouting at them, "I know you did
this and I don't want to hear any lies!" The reason they signed, they
later said, was they thought that only guilty people needed a lawyer,
whereas they had done nothing wrong and had nothing to hide. "It
appears," the experimenters concluded mournfully, "that people have
a naïve faith in the power of their own innocence to set them free."
Saul M. Kassin and Rebecca J. Norwick, "Why People Waive Their
Miranda Rights: The Power of Innocence," *Law and Human Behavior*
28 (2004): 211–21.

38. Drizin and Leo, "The Problem of False Confessions in the Post-DNA
World," 948.

39. For example, one teenager, Kharey Wise, was told that the jogger was
hit with a "very heavy object" and then was asked, "Was she hit with a
stone or brick?" Wise said first that it was a rock; moments later that
it was a brick. He said one of the others had pulled out a knife and
cut the jogger's shirt off, which wasn't true; there were no knife cuts.
Saul Kassin, "False Confessions and the Jogger Case," *New York Times,*
November 1, 2002.

40. *New York v. Kharey Wise, Kevin Richardson, Antron McCray, Yusef
Salaam, and Raymond Santana;* affirmation in response to motion to
vacate judgment of conviction, Indictment No. 4762/89, by Assistant
District Attorney Nancy Ryan, December 5, 2002, 46.

41. Gary L. Wells, "Eyewitness Identification: Probative Value, Criterion
Shifts, and Policy Regarding the Sequential Lineup," *Current Directions
in Psychological Science* 23 (2013): 11–16.

42. Adam Liptak, "Prosecutors Fight DNA Use for Exoneration," *New York Times,* August 29, 2003. See also Medwed, "The Zeal Deal," for a review of the evidence of prosecutorial resistance to reopening DNA cases. For the story of Wilton Dedge, see http://www.innocenceproject.org/cases-false-imprisonment/wilton-dedge.

43. Sara Rimer, "Convict's DNA Sways Labs, Not a Determined Prosecutor," *New York Times,* February 6, 2002.

44. "The Case for Innocence," a *Frontline* special for PBS by Ofra Bikel, first aired October 31, 2000. Transcripts and information available at http://www.pbs.org/wgbh/pages/frontline/shows/case/etc/tapes.html.

45. Drizin and Leo, "The Problem of False Confessions in the Post-DNA World," 928, footnote 200.

46. Adam Liptak, "In Appeal, Scrutiny on Not One but 3 Confessions," *New York Times,* May 20, 2014. See also www.thedailybeast.com/articles/2014/06/19/the-supreme-court-must-right-the-wrong-done-to-billy-wayne-cope.html.

47. In a famous case in North Carolina in which a victim identified the wrong man as the man who raped her, the DNA was eventually traced back to the true perpetrator; see James M. Doyle, *True Witness: Cops, Courts, Science, and the Battle Against Misidentification* (New York: Palgrave Macmillan, 2005). Sometimes, too, a cold case is solved with DNA evidence. In Los Angeles in 2004, detectives working in the newly formed cold case unit got samples of semen from the body of a woman who had been raped and murdered years earlier and checked them against the state's database of DNA from convicted violent felons. They got a match to Chester Turner, who was already in prison for rape. The detectives kept submitting DNA samples from other unsolved murders to the lab, and every month they got another match with Turner. Before long, they had linked him to twelve slayings of poor black prostitutes. Amid the general exhilaration of catching a serial killer, district attorney Steve Cooley quietly released David Jones, a janitor with severe mental retardation, who had spent nine years in prison for three of the murders.

If Turner had murdered only those three women, he would still be at large and Jones would still be in prison. But because Turner killed nine other women whose cases were unsolved, Jones was the lucky beneficiary of the efforts of the cold case unit. Justice, for him, was a byproduct of another investigation. No one, not even the cold case investigators, had any motivation to check Jones's DNA against the samples from the victims during those long nine years. But the new team of detectives had every motivation to solve old unsolved crimes, and that is the only reason that justice was served and Jones was released.

48. Deborah Davis and Richard Leo, "Strategies for Preventing False Confessions and Their Consequences," in *Practical Psychology for Forensic Investigations and Prosecutions,* eds. M. R. Kebbell and G. M. Davies (Chichester, England: Wiley, 2006), 121–49. See also the essays in Saundra D. Westervelt and John A. Humphrey, eds., *Wrongly Convicted: Perspectives on Failed Justice* (New Brunswick, NJ: Rutgers University Press, 2001); and Saul M. Kassin, "Why Confessions Trump Innocence," *American Psychologist* 67 (2012): 431–45.

49. "The Case for Innocence," *Frontline.*

50. D. Michael Risinger and Jeffrey L. Loop, "Three Card Monte, Monty Hall, Modus Operandi and 'Offender Profiling': Some Lessons of Modern Cognitive Science for the Law of Evidence," *Cardozo Law Review* 24 (November 2002): 193.

51. Mark Godsey, *Blind Justice: A Former Prosecutor Exposes the Psychology and Politics of Wrongful Convictions* (Oakland: University of California Press, 2017), 27–28.

52. Davis and Leo, "Strategies for Preventing False Confessions," 145.

53. McClurg, "Good Cop, Bad Cop." McClurg's own suggestions for using cognitive dissonance to reduce the risk of police lying are in this essay.

54. Updated statistics on the states that require electronic recording are courtesy of Rebecca Brown, Director of Policy, Innocence Project, 40 Worth Street, Suite 701, New York, NY 10013. See also the section on "Videotaping Interrogations: A Policy Whose Time Has Come," in Saul M. Kassin and Gisli H. Gudjonsson, "The Psychology of Confession

Evidence: A Review of the Literature and Issues," *Psychological Science in the Public Interest* 5 (2004): 33–67. See also Drizin and Leo, "The Problem of False Confessions in the Post-DNA World"; Davis and O'Donohue, "The Road to Perdition."

55. Quoted in Jerome, "Suspect Confession."

56. Thomas P. Sullivan, "Police Experiences with Recording Custodial Interrogations," 2004. This study, with extensive references on the benefits of recordings, is posted at http://www.law.northwestern.edu/wrongfulconvictions/Causes/custodialInterrogations.htm. Sullivan keeps a running tally of how many states are using electronic recordings; see www.nacdl.org/electronicrecordingproject. However, further research has shown that the camera angle can bias observers' judgments, especially if the camera is focused exclusively on the suspect and does not include the interviewer. G. Daniel Lassiter et al., "Videotaped Interrogations and Confessions: A Simple Change in Camera Perspective Alters Verdicts in Simulated Trials," *Journal of Applied Psychology* 87 (2002): 867–74.

57. Gisli H. Gudjonsson and John Pearse, "Suspect Interviews and False Confessions," *Current Directions in Psychological Science* 20 (2011): 33–37.

58. Davis and Leo, "Strategies for Preventing False Confessions," 145.

59. Moore and Fitzsimmons, "Justice Imperiled," 542.

60. Douglas Starr, "The Interview," *New Yorker,* December 9, 2013, 42–49. Quote is on page 49. Starr ends his piece with the story of Darrel Parker, who made a false confession to John Reid in 1955 that the courts later agreed had been coerced. In 2012, Nebraska's state attorney general publicly apologized to Parker, then eighty years old, and offered him $500,000 in damages. "Today, we are righting the wrong done to Darrel Parker more than fifty years ago," he said. "Under coercive circumstances, he confessed to a crime he did not commit." To our knowledge, John Reid has never discussed, much less apologized for, his role in Parker's false confession.

61. Thomas Vanes, "Let DNA Close Door on Doubt in Murder Cases," *Los Angeles Times,* July 28, 2003.

6. LOVE'S ASSASSIN: SELF-JUSTIFICATION IN MARRIAGE

1. John Butler Yeats to his son William, November 5, 1917, in *Letters to W. B. Yeats,* eds. Richard J. Finneran, George M. Harper, and William M. Murphy, vol. 2 (New York: Columbia University Press, 1977), 338.

2. Andrew Christensen and Neil S. Jacobson, *Reconcilable Differences* (New York: Guilford, 2000). We have taken excerpts from the story of Debra and Frank at the opening of chapter 1. This story remains in the book's updated second edition, by Andrew Christensen, Brian D. Doss, and Neil S. Jacobson, published in 2014.

3. See Neil S. Jacobson and Andrew Christensen, *Acceptance and Change in Couple Therapy: A Therapist's Guide to Transforming Relationships* (New York: W. W. Norton, 1998).

4. Christensen and Jacobson, *Reconcilable Differences,* 9.

5. There is a very large body of research on the way a couple's attributions about each other affect their feelings about each other and the course of their marriage. See Adam Davey et al., "Attributions in Marriage: Examining the Entailment Model in Dyadic Context," *Journal of Family Psychology* 15 (2001) 721–34; Thomas N. Bradbury and Frank D. Fincham, "Attributions and Behavior in Marital Interaction," *Journal of Personality and Social Psychology* 63 (1992): 613–28; and Benjamin R. Karney and Thomas N. Bradbury, "Attributions in Marriage: State or Trait? A Growth Curve Analysis," *Journal of Personality and Social Psychology* 78 (2000): 295–309.

6. June P. Tangney, "Relation of Shame and Guilt to Constructive versus Destructive Responses to Anger Across the Lifespan," *Journal of Personality and Social Psychology* 70 (1996): 797–809.

7. John Gottman, *Why Marriages Succeed or Fail* (New York: Simon and Schuster, 1994). Fred and Ingrid are on page 69.

8. Benjamin R. Karney and Thomas N. Bradbury, "The Longitudinal Course of Marital Quality and Stability: A Review of Theory, Method, and Research," *Psychological Bulletin* 118 (1995): 3–34; and Frank D. Fincham, Gordon T. Harold, and Susan Gano-Phillips, "The

Longitudinal Relation between Attributions and Marital Satisfaction: Direction of Effects and Role of Efficacy Expectations," *Journal of Family Psychology* 14 (2000): 267–85.

9. Gottman, *Why Marriages Succeed or Fail,* 57.

10. Quoted in Ayala M. Pines, "Marriage," in *Every-Woman's Emotional Well-Being,* ed. C. Tavris (New York: Doubleday, 1986), 190–91.

11. Julie Schwartz Gottman, ed., *The Marriage Clinic Casebook* (New York: W. W. Norton, 2004), 50.

12. Gottman, *Why Marriages Succeed or Fail,* 127, 128.

13. Donald T. Saposnek and Chip Rose, "The Psychology of Divorce," in *Handbook of Financial Planning for Divorce and Separation,* eds. D. L. Crumbley and N. G. Apostolou (New York: John Wiley, 1990). Their article is available online at http://www.mediate.com/articles/saporo.cfm. For a classic study of the ways that couples reconstruct their memories of their marriage and each other, see Janet R. Johnston and Linda E. Campbell, *Impasses of Divorce: The Dynamics and Resolution of Family Conflict* (New York: Free Press, 1988).

14. Jacobson and Christensen, *Acceptance and Change in Couple Therapy,* discuss new approaches to help partners accept each other rather than always trying to get the other to change.

15. Vivian Gornick, "What Independence Has Come to Mean to Me: The Pain of Solitude, the Pleasure of Self-Knowledge," in *The Bitch in the House,* ed. Cathi Hanauer (New York: William Morrow, 2002), 259.

7. WOUNDS, RIFTS, AND WARS

1. Our portrayal of this couple is based on the story of Joe and Mary Louise in Andrew Christensen and Neil S. Jacobson, *Reconcilable Differences* (New York: Guilford, 2000), 290.

2. The story of the Schiavo family battle is drawn from news reports and the in-depth reporting by Abby Goodnough, "Behind Life-and-Death Fight, a Rift that Began Years Ago," *New York Times,* March 26, 2005.

3. Sukhwinder S. Shergill et al., "Two Eyes for an Eye: The Neuroscience of Force Escalation," *Science* 301 (July 11, 2003): 187.

4. Roy F. Baumeister, Arlene Stillwell, and Sara R. Wotman, "Victim and Perpetrator Accounts of Interpersonal Conflict: Autobiographical Narratives about Anger," *Journal of Personality and Social Psychology* 59 (1990): 994–1005. The examples of typical remarks are ours, not the researchers'.

5. Timothy Garton Ash, "Europe's Bloody Hands," *Los Angeles Times,* July 27, 2006.

6. Luc Sante, "Tourists and Torturers," *New York Times,* May 11, 2004.

7. Amos Oz, "The Devil in the Details," *Los Angeles Times,* October 10, 2005.

8. Riccardo Orizio, *Talk of the Devil: Encounters with Seven Dictators* (New York: Walker and Company, 2003).

9. Louis Menand, "The Devil's Disciples: Can You Force People to Love Freedom?," *New Yorker,* July 28, 2003.

10. Keith Davis and Edward E. Jones, "Changes in Interpersonal Perception as a Means of Reducing Cognitive Dissonance," *Journal of Abnormal and Social Psychology* 61 (1960): 402–10; see also Frederick X. Gibbons and Sue B. McCoy, "Self-Esteem, Similarity, and Reactions to Active versus Passive Downward Comparison," *Journal of Personality and Social Psychology* 60 (1961): 414–24.

11. Yes, he really said it. Derrick Z. Jackson, "The Westmoreland Mind-Set," *Boston Globe,* July 20, 2005. Westmoreland made these remarks in the 1974 Vietnam documentary *Hearts and Minds.* According to Jackson, "The quote so stunned director Peter Davis that he gave Westmoreland a chance to clean it up." He didn't.

12. Ellen Berscheid, David Boye, and Elaine Walster, "Retaliation as a Means of Restoring Equity," *Journal of Personality and Social Psychology* 10 (1968): 370–76.

13. Stanley Milgram, *Obedience to Authority* (New York: Harper and Row, 1974), 10.

14. On demonizing the perpetrator as a way of restoring consonance and maintaining a belief that the world is just, see John H. Ellard et al., "Just World Processes in Demonizing," in *The Justice Motive in Everyday Life,*

eds. M. Ross and D. T. Miller (New York: Cambridge University Press, 2002).

15. John Conroy, *Unspeakable Acts, Ordinary People* (New York: Knopf, 2000), 112.

16. On December 9, 2014, the Senate Intelligence Committee released its sweeping indictment of the CIA's program to detain, interrogate, and torture terrorism suspects.

17. Cheney appeared on *Meet the Press* on December 14, 2014, http:// www.nbcnews.com/meet-the-press/meet-press-transcript-december-14-2014-n268181. See also Paul Waldman, "Why It Matters That Dick Cheney Still Can't Define Torture," *Washington Post,* December 15, 2014.

18. Bush made his remark on November 7, 2005, after news that detainees were being held in secret "terror jails" and the abuses at Abu Ghraib prison had been exposed. Inhofe made his comments on May 11, 2004, during the Senate Armed Services Committee hearings regarding abuses of Iraqi prisoners at Abu Ghraib prison. In February 2004, the International Committee of the Red Cross issued its findings, "Report of the International Committee of the Red Cross (ICRC) on the Treatment by the Coalition Forces of Prisoners of War and Other Protected Persons by the Geneva Conventions in Iraq during Arrest, Internment and Interrogation." This document is available at http:// www.globalsecurity.org/military/library/report/2004/icrc_report_ iraq_feb2004.htm. Under #1, "Treatment During Arrest," see point 7: "Certain CF [Coalition Forces] military intelligence officers told the ICRC that in their estimate between 70% and 90% of the persons deprived of their liberty in Iraq had been arrested by mistake."

19. Charles Krauthammer made the case for the limited use of torture in "The Truth about Torture: It's Time to Be Honest About Doing Terrible Things," *Weekly Standard,* December 5, 2005.

20. *New York Times* editorial, December 10, 2005, commenting on the case of Ibn al-Shaykh al-Libi, a former al-Qaeda leader who was captured in Pakistan by American forces and sent for "questioning" to Egypt. The Egyptians sent him back to the American authorities when he finally

confessed that al-Qaeda members had received chemical weapons training in Iraq — information the Americans wanted to hear. Later, Libi said he made the story up to appease the Egyptians, who were torturing him, with American approval.

21. Remarks of Condoleezza Rice at Andrews Air Force Base, December 5, 2005, as she was departing for a state visit to Europe.

22. William Schulz, "An Israeli Interrogator, and a Tale of Torture," letter to the *New York Times,* December 27, 2004.

23. An anonymous sergeant describing the handling of detainees in Iraq in a Human Rights Watch report, September 2005; reprinted with other commentary in "Under Control," *Harper's,* December 2005, 23–24.

24. For poll numbers, see the Pew Research center website, http://www. people-press.org/2014/12/15/about-half-see-cia-interrogation-methods-as-justified/. For a Pew Center review of the polls, see http://www. pewresearch.org/fact-tank/2014/12/09/americans-views-on-use-of-torture-in-fighting-terrorism-have-been-mixed/.

25. Quoted in Jane Mayer, "Torture and the Truth," *New Yorker,* December 22 and 29, 2014, 43–44. See also Antonio M. Taguba, "Stop the CIA Spin on Torture," *New York Times,* August 6, 2014. In 2004, Major General Taguba was sent to investigate the abuses at Abu Ghraib, and he'd reported systematic problems of criminal actions: "My report's findings, which prompted a Senate Armed Services Committee hearing, documented a systemic problem: military personnel had perpetrated 'numerous incidents of sadistic, blatant, and wanton criminal abuses.' The report led to prosecutions, reform of interrogation and detention regulations, and improved training. "But the military's path to accountability was a long one, and its leaders hardly welcomed oversight." In 2007, he was asked to resign.

26. Senator McCain's statement is available on his website, http://www. mccain.senate.gov/public/index.cfm/2014/12/floor-statement-by-sen-mccain-on-senate-intelligence-committee-report-on-cia-interrogation-methods.

27. Christensen and Jacobson, *Reconcilable Differences,* 291.

28. For a thoughtful analysis of the social and personal costs of forgiveness that is uncritical and premature, letting perpetrators off the hook of responsibility and accountability for the harm they caused, see Sharon Lamb, *The Trouble with Blame: Victims, Perpetrators, and Responsibility* (Cambridge, MA: Harvard University Press, 1996).

29. Solomon Schimmel, *Wounds Not Healed by Time: The Power of Repentance and Forgiveness* (Oxford, England: Oxford University Press, 2002), 226. Psychologist Ervin Staub, himself a Holocaust survivor, has been studying the origins and dynamics of genocide for many years and has devoted himself to the project of reconciliation between the Tutsi and Hutu in Rwanda. See Ervin Staub and Laurie A. Pearlman, "Advancing Healing and Reconciliation in Rwanda and Other Post-conflict Settings," in *Psychological Interventions in Times of Crisis,* eds. L. Barbanel and R. Sternberg (New York: Springer-Verlag, 2006); and Daniel Goleman, *Social Intelligence* (New York: Bantam Books, 2006).

30. Broyles told this story in a May 27, 1987, PBS documentary *Faces of the Enemy,* based on the book of the same title by Sam Keen.

8. LETTING GO AND OWNING UP

1. Wayne Klug et al., "The Burden of Combat: Cognitive Dissonance in Iraq War Veterans," in *Treating Young Veterans,* eds. Diann C. Kelly et al. (New York: Springer, 2011), 33–80.

2. Dexter Filkins, "Atonement: A Troubled Iraq Veteran Seeks Out the Family He Harmed," *New Yorker,* October 29 and November 5, 2012, 92–103.

3. Nell Greenfieldboyce, "Wayne Hale's Insider's Guide to NASA," NPR *Morning Edition,* June 30, 2006.

4. Jennifer K. Robbennolt, "Apologies and Settlement Levers," *Journal of Empirical Legal Studies* 3 (2008): 333–73.

5. Cass Sunstein, "In Politics, Apologies Are for Losers," *New York Times,* July 27, 2019, https://www.nytimes.com/2019/07/27/opinion/sunday/when-should-a-politician-apologize.html.

6. Ronald Reagan perfected the apology-without-its-essence language

in his response to the Iran-Contra scandal of the mid-1980s, in which administration officials secretly arranged an illegal sale of arms to Iran and used the money to fund the Contras in Nicaragua. Reagan's defense started out well — "First, let me say I take full responsibility for my own actions and for those of my administration" — but then he added a series of "but they did it"s: "As angry as I may be about activities undertaken without my knowledge, I am still accountable for those activities. As disappointed as I may be in some who served me, I'm still the one who must answer to the American people for this behavior. And as personally distasteful as I find secret bank accounts and diverted funds — well, as the Navy would say, this happened on my watch." And this is how he took "full responsibility" for breaking the law: "A few months ago I told the American people I did not trade arms for hostages. My heart and my best intentions still tell me that's true, but the facts and the evidence tell me it is not."

7. Lisa Leopold, *The Conversation*, February 8, 2019, https://theconversation.com/how-to-say-im-sorry-whether-youve-appeared-in-a-racist-photo-harassed-women-or-just-plain-screwed-up-107678.

8. Daniel Yankelovich and Isabella Furth, "The Role of Colleges in an Era of Mistrust," *Chronicle of Higher Education* (September 16, 2005): B8–B11.

9. Posted on the website of an advocacy group called Sorry Works!, a coalition of physicians, hospital administrators, insurers, patients, and others concerned with the medical-malpractice crisis. See also Katherine Mangan, "Acting Sick," *Chronicle of Higher Education* (September 15, 2006), and Robbennolt, "Apologies and Settlement Levers."

10. Atul Gawande, *Being Mortal* (New York: Henry Holt, 2014). See also "The Problem of Hubris," the third of his fourth 2014 Reith lectures recorded on BBC4: http://www.bbc.co.uk/programmes/articles/6F2X8TpsxrJpnsq82hggHW/dr-atul-gawande-2014-reith-lectures.

11. Richard A. Friedman, "Learning Words They Rarely Teach in Medical School: 'I'm Sorry,'" *New York Times,* July 26, 2005.

12. Warren G. Bennis and Burt Nanus, *Leaders: Strategies for Taking Charge,* rev. ed. (New York: HarperCollins, 1995), 70.

13. Atul Gawande, *The Checklist Manifesto: How to Get Things Right* (New York: Henry Holt, 2009).

14. *Harmful Error: Investigating America's Local Prosecutors,* Center for Public Integrity (Summer 2003), http://www.publicintegrity.org.

15. Meytal Nasie et al., "Overcoming the Barrier of Narrative Adherence in Conflicts Through Awareness of the Psychological Bias of Naïve Realism," *Personality and Social Psychology Bulletin* 40 (2014): 1543–56.

16. Quoted in Dennis Prager's *Ultimate Issues* (Summer 1985): 11.

17. Joe Coscarelli, "Michael Jackson Fans Are Tenacious. 'Leaving Neverland' Has Them Poised for Battle," *New York Times,* March 4, 2019, https://www.nytimes.com/2019/03/04/arts/music/michael -jackson-leaving-neverland-fans.html?searchResultPosition=2.

18. Amanda Petrusich, "A Day of Reckoning for Michael Jackson with 'Leaving Neverland,'" *New Yorker,* March 1, 2019.

19. Margo Jefferson, introduction to new edition of *On Michael Jackson* (New York: Penguin, 2019).

20. Anthony Pratkanis and Doug Shadel, *Weapons of Fraud: A Source Book for Fraud Fighters* (Seattle, WA: AARP, 2005).

21. Stigler recalled this story in an obituary for Harold Stevenson, the *Los Angeles Times,* July 22, 2005. For their research, see Harold W. Stevenson and James W. Stigler, *The Learning Gap* (New York: Summit, 1992); and Harold W. Stevenson, Chuansheng Chen, and Shin-ying Lee, "Mathematics Achievement of Chinese, Japanese, and American Schoolchildren: Ten Years Later," *Science* 259 (January 1, 1993): 53–58.

22. Carol S. Dweck, "The Study of Goals in Psychology," *Psychological Science* 3 (1992): 165–67; Claudia M. Mueller and Carol S. Dweck, "Praise for Intelligence Can Undermine Children's Motivation and Performance," *Journal of Personality and Social Psychology* 75 (1998): 33–52.

23. Hampton Stevens has also made this point. "The idea that Fitzgerald, of all people, didn't believe Americans could reinvent themselves is a like thinking Tolstoy didn't believe in snow," he wrote. See "Why Tiger Woods Isn't Getting a 'Second Act,'" *Atlantic,* April 2010.

24. Laura A. King and Joshua A. Hicks, "Whatever Happened to 'What Might Have Been'? Regrets, Happiness, and Maturity," *American Psychologist* 62 (2007): 625–36.

25. Matt Richtel, "A Texting Driver's Education," *New York Times,* September 13, 2014. Richtel is also the author of *A Deadly Wandering: A Tale of Tragedy and Redemption in the Age of Attention* (New York: William Morrow, 2014). As of 2019, Reggie Shaw's website still lists his lectures, volunteering, and other work to promote driving safety.

26. Eric Fair, "I Can't Be Forgiven for Abu Ghraib," *New York Times,* December 10, 2014.

27. Tina Nguyen, "Fox's Andrea Tantaros Dismisses Torture Report Because 'America Is Awesome,'" December 9, 2014, http://www.mediaite.com/tv/foxs-andrea-tantaros-dismisses-torture-report-because-america-is-awesome/.

28. Quoted in Charles Baxter, "Dysfunctional Narratives, or: 'Mistakes Were Made,'" in *Burning Down the House: Essays on Fiction* (Saint Paul, MN: Graywolf Press, 1997), 5. There is some dispute about the second sentence of Lee's remarks but not about his assuming responsibility for the disastrous results of his military decisions.

29. Daniel Bolger, "Why We Lost in Iraq and Afghanistan," *Harper's,* September 2014, 63–65.

30. Eisenhower's handwritten document is available at http://www.archives.gov/education/lessons/d-day. See also Michael Korda, *Ike: An American Hero* (New York: HarperCollins, 2007). On the *Charlie Rose* show on November 16, 2007, Korda said of Eisenhower: "When things went right, he praised his subordinates and made sure they got the praise; and when things went wrong, he took the blame. Not many presidents have done that, and very few generals."

9. DISSONANCE, DEMOCRACY, AND THE DEMAGOGUE

1. David I. Kertzer, *The Pope and Mussolini: The Secret History of Pius XI and the Rise of Fascism in Europe* (New York: Random House, 2014).

2. Ibid., 29.

3. Ibid., 56.

4. Peter Eisner, *The Pope's Last Crusade: How an American Jesuit Helped Pope Pius XI's Campaign to Stop Hitler* (New York: William Morrow, 2013), 51.

5. https://www.nbcnews.com/think/opinion/trump-s-presidency-was -made-possible-historical-demagogues-joe-mccarthy-ncna817981.

6. Peter Baker and Michael D. Shear, "El Paso Shooting Suspect's Manifesto Echoes Trump's Language," *New York Times,* August 4, 2019. Other reporters and organizations have been keeping track of the "Trump supporters, fans, and sympathizers [who] have beaten, shot, stabbed, run over, and bombed their fellow Americans . . . while aping the president's violent rhetoric"; see https://theintercept.com/2018/10/ 27/here-is-a-list-of-far-right-attackers-trump-inspired-cesar-sayoc-wasnt -the-first-and-wont-be-the-last/.

7. https://qz.com/1307928/fire-and-fury-author-michael-wolff-breaks -down-donald-trumps-sales-tactics/.

8. Emily Ekins, "The Five Types of Trump Voters: Who They Are and What They Believe," Research Report from the Democracy Fund Voter Study Group, Pew Research Center, June 19, 2017.

9. https://www.washingtonpost.com/opinions/ari-fleischer-heres-how-i -figured-out-whom-to-vote-for/2016/11/04/7bcee1ec-a1fd-11e6-8d63 -3e0a660f1f04_story.html?utm_term=.1a26e9d00616. See also Isaac Chotiner, "Ari Fleischer on Why Former Republican Critics of Trump Now Embrace Him," *New Yorker,* July 9, 2019.

10. See https://www.businessinsider.com/trump-cruz-feud-history-worst -attacks-2016-9#trump-the-state-of-iowa-should-disqualify-ted-cruz

-from-the-most-recent-election-on-the-basis-that-he-cheated-a-total
-fraud-11.

Cruz's quote about holding Mussolini's jacket is from Tim Alberta, *American Carnage: On the Front Lines of the Republican Civil War and the Rise of President Trump* (New York: Harper, 2019). See also https:// www.washingtonpost.com/politics/new-book-details-how-republican -leaders-learned-to-stop-worrying-and-love-trump/2019/07/10/ be75eff8-a27d-11e9-b7b4-95e30869bd15_story.html.

11. Lindsey Graham, CNN interview, December 8, 2015, https://www .mcclatchydc.com/news/politics-government/election/article62680527 .html#storylink=cpy.

12. Jonathan Freedland, "Anti-Vaxxers, the Momo Challenge . . . Why Lies Spread Faster Than Facts," *Guardian,* March 8, 2019, https:// www.theguardian.com/books/2019/mar/08/anti-vaxxers-the-momo -challenge-why-lies-spread-faster-than-facts?CMP=share_btn_link.

13. Zachary Jonathan Jacobson, "Many Are Worried About the Return of the 'Big Lie.' They're Worried About the Wrong Thing," *Washington Post,* May 21, 2018.

14. Aaron Rupar, "Trump's Bizarre 'Tim/Apple' Tweet Is a Reminder the President Refuses to Own Tiny Mistakes," *Vox,* March 11, 2019.

15. Glenn Kessler, Salvador Rizzo, and Meg Kelly, "President Trump Has Made 15,413 False or Misleading Claims Over 1,055 Days," *Washington Post,* December 16, 2019, https://www.washingtonpost.com/politics/ 2019/12/16/president-trump-has-made-false-or-misleading-claims-over -days/?smid=nytcore-ios-share.

16. At an infamous July 18, 2019, rally in which he launched into an attack on Omar, his fans began shouting "Send her back!" Trump stood at his podium, passively absorbing their chant for a full thirteen seconds, before moving on. Yet when he was criticized later for not stopping the mob's nasty refrain, he said he "disagreed with it" but there was nothing he could do, so "I started speaking very quickly." The videotape shows that he did not.

17. Justin Baragona, "Fox Business Host Stuart Varney Tells Joe Walsh: Trump Has Never Lied," *Daily Beast,* August 30, 2019.

18. The list of people who had resigned or been fired as of June 2019 can be found at https://www.businessinsider.com/who-has-trump-fired-so-far -james-comey-sean-spicer-michael-flynn-2017-7.

19. See https://qz.com/1267508/all-the-people-close-to-donald-trump-who -called-him-an-idiot/;
    Annalisa Merelli and Max de Haldevang, "All the Ways Trump's Closest Confidants Insult His Intelligence," *Quartz,* May 2, 2018.

20. From the beginning of Trump's presidency, his mental stability has been called into question, specifically his narcissism and grandiosity, his detachment from reality, his erratic behavior, and his rages. See George T. Conway III, "Unfit for Office," *Atlantic,* October 3, 2019.

21. Anonymous, "I Am Part of the Resistance Inside the Trump Administration," *New York Times,* September 5, 2018.

22. Eric Levitz, "Mueller Report Confirms Trump Runs the White House Like It's the Mafia," *New York,* April 18, 2019. See also Jonathan Chait, "Trump Wants to Ban Flipping Because He Is Almost Literally a Mob Boss," *New York,* August 23, 2018.

23. Kevin Liptak, "Trump Says Longstanding Legal Practice of Flipping 'Almost Ought to Be Illegal,'" CNN, August 23, 2018, https://www.cnn .com/2018/08/23/politics/trump-flipping-outlawed/index.html.

24. Isaac Chotiner, "Ari Fleischer on Why Former Republican Critics of Trump Now Embrace Him," *New Yorker,* July 9, 2019.

25. Lisa Lerer and Elizabeth Dias, "Israel's Alliance with Trump Creates New Tensions Among American Jews," *New York Times,* August 17, 2019, https://www.nytimes.com/2019/08/17/us/politics/trump-israel-jews .html.

26. Peter Wehner, "What I've Gained by Leaving the Republican Party," *Atlantic,* February 6, 2019, https://www.theatlantic.com/ideas/archive/ 2019/02/i-left-gop-because-trump/581965/.

27. Stephanie Denning, "Why Won't the Trump Administration Admit a Mistake?," *Forbes,* March 17, 2018.

28. Greg Weiner, "The Trump Fallacy," *New York Times*, July 1, 2019.

29. James Comey, "How Trump Co-Opts Leaders Like Barr," *New York Times*, May 1, 2019. Comey tells of his disgust at Barr's "Congratulations, Mr. President" remark.

30. Jamelle Bouie, "The Joy of Hatred," *New York Times,* July 19, 2019, https://www.nytimes.com/2019/07/19/opinion/trump-rally.html.

31. Adam Gopnik, "Europe and America Seventy-Five years after D-Day," *New Yorker,* June 6, 2019. The only thing more alarming than Trump's assault on the principles and practices of liberal democracy, Gopnik wrote, is "the ease with which his actions have been normalized and treated as eccentricities rather than the affronts to liberal democratic values that, for all their seeming triviality, they are. Principles are built out of many bricks; even the loss of one weakens the whole."

32. Peter Nicholas, "It Makes Us Want to Support Him More," *Atlantic,* July 18, 2019.

33. Julie Hirschfeld Davis and Katie Rogers, "At Trump Rallies, Women See a Hero Protecting Their Way of Life," *New York Times,* November 3, 2018.

34. Ashley Jardina, interview by Chauncey DeVega, *Salon,* July 17, 2019, https://www.salon.com/2019/07/17/author-of-white-identity-politics -we-really-need-to-start-worrying-as-a-country/. See also Ashley Jardina, *White Identity Politics* (Cambridge: Cambridge University Press, 2019).

35. See https://www.theatlantic.com/politics/archive/2019/10/trump -white-evangelical-impeachment/600376/?utm_source=atl&utm _medium=email&utm_campaign=share.

36. See https://www.minnpost.com/eric-black-ink/2016/06/donald-trump -s-breathtaking-self-admiration/.

37. Hemant Mehta, "Contradicting Herself, Christian Says Morality Is Now Optional for a President," *Friendly Atheist,* July 27, 2019; https:// friendlyatheist.patheos.com/2019/07/27/contradicting-herself-christian -says-morality-is-now-optional-for-a-president/?utm_source=dlvr.it& utm_medium=facebook.

38. Susan B. Glasser, "Mike Pompeo, the Secretary of Trump," *New Yorker,*

August 26, 2019. On March 21, 2019, Pompeo was interviewed on the Christian Broadcasting Network, where he was reminded of Esther's role in persuading the Persian king not to follow Haman's evil plan to eradicate the Jews. "Could it be that President Trump right now has been sort of raised for such a time as this, just like Queen Esther, to help save the Jewish people from the Iranian menace?" "As a Christian, I certainly believe that's possible," replied Pompeo. See https://www.washingtonpost.com/opinions/this-is-trumps-year-of-living-biblically/2019/03/27/e3d00802-50c9-11e9-8d28-f5149e5a2fda_story.html.

39. https://beta.washingtonpost.com/politics/trump-uttered-what-many-supporters-consider-blasphemy-heres-why-most-will-probably-forgive-him/2019/09/13/685c0bce-d64f-11e9-9343-40db57cf6abd_story.html.

40. Allen and Reed both quoted in Tom McCarthy, "Faith and Freedoms: Why Evangelicals Profess Unwavering Love for Trump," *Guardian,* July 7, 2019; https://www.theguardian.com/us-news/2019/jul/07/donald-trump-evangelical-supporters?CMP=share_btn_link

41. Max Boot, *The Corrosion of Conservatism: Why I Left the Right* (New York: Norton, 2018), xxi, 58.

42. Wehner, "What I've Gained by Leaving the Republican Party." What did he gain? "I'm more willing to listen to those I once thought didn't have much to teach me" — a dissonance lesson for all of us.

43. Ben Howe, *The Immoral Majority: Why Evangelicals Chose Political Power Over Christian Values* (New York: Broadside, 2019), 170.

44. Laurie Goodstein, "'This Is Not of God': When Anti-Trump Evangelicals Confront Their Brethren," *New York Times,* May 23, 2018.

45. Maria Caffrey, "I'm a Scientist. Under Trump I Lost My Job for Refusing to Hide Climate Crisis Facts," *Guardian,* July 25, 2019.

46. Chuck Park, "I Can No Longer Justify Being a Part of Trump's 'Complacent State.' So I'm Resigning," *Washington Post,* August 8, 2019; Bethany Milton, *New York Times,* August 26, 2019, wrote a similar piece explaining her resignation from the State Department.

47. Sharon La Franiere, Nicholas Fandos, and Andrew E. Kramer, "Ukraine

Envoy Says She Was Told Trump Wanted Her Out Over Lack of Trust," *New York Times,* October 11, 2019.

48. David Remnick, "Trump Clarification Syndrome," *New Yorker,* August 23, 2019.

49. Jim Mattis and Bing West, *Call Sign Chaos: Learning to Lead* (New York: Random House, 2019).

INDEX

Littwin, Mike, 77–78
lobbyists, 61–62
lockjaw, 9, 33
Loftus, Elizabeth, 119
logical fallacies, 151, 355
Long, Huey, 333
Loop, Jeffrey L., 208
lost possible selves, 315–17
Luhrman, Tanya, 140
Lundberg, Ferdinand, 10
lying
    self-justification versus, 4–5
    by Trump, 341–45

Mack, John, 126
Maechler, Stefan, 116, 118
magazine-subscription scams, 306
magic ratio, 235–36
Magruder, Jeb Stuart, 48–51, 52–53, 93–94, 355, 368
Mandela, Nelson, 282
Marino, Gordon, 6
Markovic, Mira, 269
Marquis, Joshua, 180, 181–82, 193
marriage and relationships
    affairs in, 249–50
    case study of, 220–34
    dissonance reduction in, 217–20
    divorce, 238–43, 269–70, 312
    empathy and, 243–44
    letting go of self-justification in, 243–47
    unhappiness in, 234–43
Marsh, Elizabeth, 106
Masten, Ann, 406–7n24
math gap, 308–9
Matta, Monika, 118, 120
Mattis, Jim, 369–70, 377
maturity, 316
Maxwell, William, 95
Mayes, Larry, 216

Mazzoni, Giuliana, 119, 400n32
McCain, John, 279, 343
McCarthy, Joseph, 333
McCarthy, Mary, 102–3
McClurg, Andrew, 189–90, 210–11
McCready, K. James, 416n28
McDonald's, 2
McDonough, Chris, 184–85, 187
McDougal, Michael, 205
McGahn, Don, 348
McMartin Preschool, 134, 137, 304
McNally, Richard J., 121–22, 126, 152–53, 169
measles, increase in cases of, 71, 72
Meehl, Paul, 143, 408n30
Meese, Edward, 196–97
Mein Kampf (Hitler), 343
Memories of a Catholic Girlhood (McCarthy), 102–3
memory
    alcohol and, 110
    biases of, 99–114
    current self-concept and, 106–8, 111–12
    distortions of, 7, 112–13
    examples of false, 114–29
    faulty nature of, 95–99, 108
    metaphors for, 100
    narrative structure of, 105–6
    nature of, 100–101
    questioning, 129–30
    reconstructing, 101–2
    third-person perspective in, 112
Menand, Louis, 269
mental-health professionals
    "believing is seeing" and, 158–64
    benevolent-dolphin problem and, 148–58
    differences among, 139–41
    dissonance reduction by, 169–71

Made in the USA
Monee, IL
16 December 2022